Machine Intelligence

Machines are being systematically empowered to be interactive and intelligent in their operations, offerings, and outputs. There are pioneering Artificial Intelligence (AI) technologies and tools. Machine and Deep Learning (ML/DL) algorithms, along with their enabling frameworks, libraries, and specialized accelerators, find particularly useful applications in computer and machine vision, human-machine interfaces (HMIs), and intelligent machines. Machines that can see and perceive can offer deeper and decisive acceleration, automation, and augmentation capabilities to businesses as well as to people in their everyday tasks. Machine vision is becoming a reality because of advances in the computer vision and device instrumentation fields. Machines are increasingly software-defined. That is, vision-enabling software and hardware modules are being embedded in new-generation machines to be self-, surroundings, and situation-aware.

Machine Intelligence emphasizes computer vision and natural language processing as drivers of advances in machine intelligence. The book examines these technologies from the algorithmic level to the applications level. It also examines the integrative technologies enabling intelligent applications in business and industry.

Features:

- Motion images object detection over voice using deep learning algorithms
- Ubiquitous computing and augmented reality in HCI
- Learning and reasoning using Artificial Intelligence
- Economic sustainability, mindfulness, and diversity in the age of artificial intelligence and machine learning
- Streaming analytics for healthcare and retail domains

Covering established and emerging technologies in machine vision, the book focuses on recent and novel applications and discusses state-of-the-art technologies and tools.

Pethuru Raj is the chief architect in the Site Reliability Engineering (SRE) division of Reliance Jio Platforms Ltd., Bangalore, India. He previously worked as a cloud infrastructure architect in the IBM Global Cloud Center of Excellence (CoE), as a TOGAF-certified enterprise architecture (EA) consultant in Wipro Consulting Services (WCS) Division and as a lead architect in the corporate research (CR) division of Robert Bosch. In total, he has more than 19 years of IT industry experience and 8 years of research experience.

P. Beaulah Soundarabai is an associate professor in the Department of Computer Science, Christ University, Bangalore, India. She has 20 years of teaching experience. She has been associated with Christ University for the past 14 years. Prior to this, she has teaching experience in SFR College for Women, Sivakasi, and as a lecturer, in AGCS, Kolkata, for 3 years respectively. She has 10 years of research experience in the areas of Distributed Computing, Computer Networks, IoT, Edge and Cloud computing, and Data Analytics.

D. Peter Augustine is an associate professor in the Department of Computer Science, Christ University, Bangalore, India. Dr. Augustine has a PhD in Medical Image Processing in Cloud Environment, with over 8 years in cloud computing and 5 years in Big Data Analytics. He has authored various research papers, that were published in peer-reviewed journals. He has been involved in a major research project using Cloud computing which costs more than 18 lakhs. He has also collaborated with St. John's Medical Research Institute on the research project to diagnose lung diseases using cutting-edge AI and Machine Learning.

Machine Intelligence

Computer Vision and Natural Language Processing

Edited by
Pethuru Raj, P. Beaulah Soundarabai, and
D. Peter Augustine

CRC Press
Taylor & Francis Group
Boca Raton London New York

CRC Press is an imprint of the
Taylor & Francis Group, an **informa** business

First edition published 2024
by CRC Press
2385 Executive Center Drive, Suite 320, Boca Raton, FL 33431

and by CRC Press
4 Park Square, Milton Park, Abingdon, Oxon, OX14 4RN

CRC Press is an imprint of Taylor & Francis Group, LLC

ISBN: 9781032201993 (hbk)
ISBN: 9781032543727 (pbk)
ISBN: 9781003424550 (ebk)

DOI: 10.1201/9781003424550

Typeset in Garamond
by Newgen Publishing UK

Contents

Figures

Tables

Preface

At the outset, the following is not an exaggerated statement:

> By smartly leveraging a growing array of digitization and digitalization technologies and tools, machines and devices in our everyday environments (homes, hotels, hospitals, retail stores, factory floors, airports, warehouses, etc.) are being systematically empowered to be interactive and intelligent in their operations, offerings and outputs.

There are path-breaking edge and digitization technologies such as miniaturized and disappearing sensors, actuators, RFID tags, stickers, microcontrollers, chips and codes, LED lights and beacons, communication modules, smart dust, etc. for transforming all kinds of physical, mechanical, electrical and electronics systems into digitized elements. Through short- and long-range communication technologies, digitized entities are subsequently connected with one another in the vicinity and with remotely held software and databases (cloud-hosted). Thus, sensing, perception, vision, communication, and computing technologies empower tangible things to become digitized entities. Such empowered entities are now ready to join in the mainstream computing. In short, all sorts of wearables, handhelds, and implantable, mobile, wireless, fixed and nomadic devices, appliances, equipment, machinery, wares, instruments, etc. are being meticulously prepared for the ensuing digital era. In the recent past, these networked embedded systems have been termed the Internet of Things (IoT) sensors and devices.

Now these connected devices, which are projected to be in billions, are capable of creating zettabytes of digital data daily. Fortunately, the IT industry is being bombarded with a variety of integrated data analytics platforms, methods, frameworks, software libraries, toolsets, and other enablers to transition digital data into actionable insights. Furthermore, data integration and virtualization techniques are prevalent, along with data fabrics and meshes to capture, aggregate and insert into data analytics platforms. The aspect of knowledge visualization is fulfilled through 360 degree dashboards. Especially with a host of breakthrough artificial intelligence (AI) algorithms and models, the difficult process of transforming digital data into insights is being hugely simplified and speeded up. By feeding the knowledge discovered into IoT devices, the days of having cognitive devices are at hand. Such edge devices can exhibit intelligent behavior.

Machine and deep learning (ML/DL) algorithms along with a host of enabling frameworks, pre-trained models, and specialized accelerators come in handy in providing computer vision (CV) and natural language processing (NLP) applications. These unique capabilities are being replicated in resource-intensive IoT devices and machines. Machines that can see and perceive can bring forth deeper and decisive acceleration, automation, and augmentation capabilities to businesses as well as people in their everyday assignments and engagements. Machine vision is becoming the grandiose reality with the distinct advances in computer vision and device instrumentation spaces. Machines are increasingly software-defined. That is, AI-enabled machines are being deployed in mission-critical locations to minutely monitor everything that is happening there and to take decisive actions in time with all the clarity and alacrity.

These pioneering developments have laid down a stimulating and scintillating foundation to visualize and realize a variety of sophisticated applications not only for enterprising businesses but also for people at large. Empowering machines with sensing, vision, perception, analytics, knowledge discovery and dissemination, decision-making and action capabilities through a host of breakthrough digital technologies and tools goes a long way in establishing and sustaining intelligent systems and societies. In an industrial setting, these technologies in conjunction with high-speed and adaptive networking, such as 5G and Wi-Fi 6, can lead to a new industrial revolution. Edge AI-inspired robots, drones, gizmos, and gadgets are bound to facilitate smart manufacturing and Industry 4.0 and 5.0 applications. Similarly the next-generation digitally transformed healthcare domain is set to immensely benefit society through such vision-enabled machines. Technology-induced innovations can aid experts and workers in conceiving and concretizing fresh ways and means to perform low-waste and high-efficiency industrial activities.

Machines are enabled to be interactive with other machines in the vicinity and with remote machines through a host of networking options. Similarly, through natural interfaces, machines and men are able to interact purposefully. That is, human-machine interfaces (HMIs) are seeing the grand reality these days with a plethora of technological innovations and disruptions. Just as convolutional neural networks (CNNs) facilitate machine vision, the faster maturity and stability of recurrent neural networks (RNNs), a grandiose part of the deep learning (DL) field, have strengthened the NLP capability of our everyday machines.

The third and final feature is to produce autonomous machines with self-configuring, healing, defending, and managing features. With the steadily growing power of machine learning (ML) algorithms, a dazzling array of industrial machinery, medical instruments, defense equipment, consumer electronics, handhelds, wearables, and other edge devices are being readied to intrinsically exhibit intelligent behavior. Devices are becoming autonomous; therefore, they can continuously and consistently deliver their designated services without any slowdown or breakdown. In their actions and reactions, they become fault-tolerant, high-performing, self-defending, adaptive, reactive, proactive, etc. In a nutshell, people and businesses will get state-of-the-art products, solutions, and services through a bevy of cutting-edge technologies.

It is possible to produce and supply a host of people-centric, real-time, service-oriented, event-driven, and context-aware services with intelligent machines in and around us. Thus, connected devices, digitized entities, and intelligent applications, the business-driven IT is all set to be people IT in the years to come.

This book is to enlighten our esteemed readers with all the details on the following topics.

1. Artificial intelligence (AI) (machine and deep learning (ML/DL) algorithms, computer vision (CV), and natural language processing (NLP) capabilities)
2. Edge computing, clouds and analytics
3. Edge AI technologies and tools
4. Industry 4.0 – a detailed overview and use cases
5. The IoT – the promising industrial applications
6. Audio and video analytics
7. Edge clouds and analytics
8. Autonomic systems for industrial processes
9. Hyperautomation techniques
10. Computer and machine vision
11. Human-machine interfaces (HMIs)
12. Cybersecurity for smart industrial systems
13. Machine vision applications and use cases for Industry 4.0
14. Smart and high-precision manufacturing
15. Intelligent robots and drones

Contributors

D. Peter Augustine
Department of Computer Science
Christ University
Bangalore, India

P. K. Nizar Banu
Department of Computer Science
Christ University
Bangalore, India

J. Yasmin Banu
Department of Computer Science
University of Madras, Chennai, India

Vaibhav Bhatnagar
Department of Computer Applications
Manipal University
Jaipur, Rajasthan, India

P. Bhavani
Department of Computer Science
University of Madras, Chennai, India

P. L. Chithra
Department of Computer Science
University of Madras, Chennai, India

D. Ruban Christoper
Department of Data Science
Loyola College, Chennai, India

Soumya Chuabey
Department of Computer Applications
Manipal University
Jaipur, Rajasthan, India

Nancy Jasmine Goldena
Department of Computer Applications
    and Research Centre
Sarah Tucker College
Tirunelveli, Tamil Nadu, India

Nishu Gupta
Department of Computer Applications
Manipal University
Jaipur, Rajasthan, India

Luthuful Haq
Department of Data Science
Loyola College, Chennai, India

P. Haritha
Department of Computer Science and
    Applications
The Gandhigram Rural Institute
Gandhigram, Dindigul
Tamil Nadu, India

Deepa V. Jose
Department of Computer Science
Christ University
Bangalore, India

R. Jyothsna
Department of Computer Science
Christ University
Bangalore, Karnataka, India

T. Kalaiselvi
Department of Computer Science and
  Applications
The Gandhigram Rural Institute
Gandhigram, Dindigul, Tamil
  Nadu, India

P. Manikandan
Department of Data Science
Loyola College, Chennai, India

G. Nagarajan
School of Computing Science and
  Engineering
Galgotias University
Greater Noida, Uttar Pradesh, India

Ardhendu G. Pathak
R&D
Airbus, GE
Bangalore, Karnataka, India

Ity Patni
Department of Computer Applications
Manipal University
Jaipur, Rajasthan, India

Joy Paulose
Department of Computer Science
Christ University
Bangalore, Karnataka, India

T. Poongodi
School of Computing Science and
  Engineering
Galgotias University
Greater Noida, Uttar Pradesh, India

Ramesh Chandra Poonia
Department of Computer Science
Christ University
Bangalore, Karnataka, India

Nidhin Raju
Department of Computer
  Science
Christ University
Bangalore, India

Sanjay A. Gobi Ramasamy
Christ University
Bangalore, India

M. Mary Shanthi Rani
Department of Computer Science and
  Applications
The Gandhigram Rural Institute
Gandhigram, Dindigul, Tamil
  Nadu, India

Merjulah Roby
Department of Mechanical
  Engineering, Vascular Biomechanics
  and Biofluids
University of Texas
San Antonio, Texas, USA

V. Rohini
Department of Computer
  Science
Christ University
Bangalore, Karnataka, India

P. Shanmugavadivu
Department of Computer Science and
  Applications
The Gandhigram Rural Institute
Gandhigram, Dindigul, Tamil
  Nadu, India

Kiran Singh
School of Computing Science and
    Engineering
Galgotias University
Greater Noida, Uttar Pradesh, India

Ranjit Singha
Department of Psychology
Christ University
Bangalore, Karnataka, India

Surjit Singha
Department of Commerce
Christ University
and
Department of Commerce
Kristu Jayanti College
Bangalore, Karnataka, India

P. Beaulah Soundarabai
Department of Computer Science
Christ University
Bangalore, India

Yadukrishna Sreekumar
Department of Computer Science
Christ University
Bangalore, India

Thangapriya
Department of Computer Applications
    and Research Centre
Sarah Tucker College
Tirunelveli, Tamil Nadu, India

Priya Thomas
Department of Computer Science
Christ University
Bangalore, India

Suman Avdhesh Yadav
Amity School of Engineering and
    Technology
Amity University
Greater Noida, Uttar Pradesh, India

Abbreviations

ACF	autocorrelation function
ADC	analog to digital converter
ADF	Augmented Dickey-Fuller
AI	artificial intelligence
ANOVA	analysis of variance
AP	average precision
API	application programming interface
AR	augmented reality
ARIMA	Auto Regressive Integrated Moving Average
ASR	Automatic Speech Recognition
BN	batch normalization
BSE	Bombay Stock Exchange
CCTV	closed-circuit television
CDS	computerized decision support
CDSSs	clinical decision support systems
CNN	convolutional neural network
CNN-LSTM	CNN Long Short-Term Memory Network
CoAP	Constrained Application Protocol
CSR	corporate social responsibility
DAC	Digital to Analog Converter
DAE	Denoising Auto Encoders
DCNN	Deep Convolutional Neural Networks
DL	deep learning
DNA	deoxyribonucleic acid
DNN	deep neural network
DNS	Domain Name Server
DPM	deformable part-based model
DSCI	Data Security Council of India
DT	Decision Tree
EDI	electronic data interchange
EGC	early gastric cancer
EHR	electronic health record

EIS	Executive Information Systems
ERP	enterprise resource planning
ETL	extract, transform, and load
FDI	foreign direct investment
FIPPs	Fair Information Practices Principles
FMCG	fast moving consumer goods
fMRI	functional Magnetic Resonance Imaging
GAN	Generative Adversarial Networks
GBDT	gradient boosting decision tree
GMM	Gaussian mixture model
GUI	graphical user interface
GWAS	Genome Wide Association Studies
HGP	Human Genome Project
HTTP	Hypertext Transmission Protocol
HUD	heads-up display
HVAC	heating, ventilation and air conditioning
IaaS	Infrastructure as a Service
ICU	intensive care unit
IDE	integrated development environment
IIoT	Industrial Internet of Things
IoT	Internet of Things
IoU	Intersection over Union
IP	Internet Protocol
IPR	intellectual property right
IR	infrared
IVAS	integrated visual augmentation system
KNN	K-Nearest Neighbor
LAN	local area network
LISP	Locator ID Separation Protocol
LLM	Large Language Model
LWAs	lightweight architectures
M2M	machine-to-machine
mAP	mean average precision
MEG	magneto encephalography
ML	machine learning
M-NBI	magnifying-narrow band Imaging
MoU	Memorandum of Understanding
MQTT	Message Queuing Telemetry Transport
MSME	micro, small and medium enterprise
NB	Naïve Bayes
NDCP	National Digital Communications Policy
NGO	non-governmental organization

NGS	Next-Generation Sequencing
NIH	National Institutes of Health
NLP	Natural Language Processing
NLR	Neutrophil-to-Lymphocyte
Nltk	Natural Language Toolkit
NMS	non-maximum suppression
NPA	non-performing asset
NPL	non-performing loan
OPLS-DA	orthogonal signal correlation-partial least square discriminant analysis
OSN	online social network
PaaS	Platform as a Service
PACF	partial autocorrelation function
PACS	picture archiving and communication system
PAN	Path Aggregation Network
PARC	Palo Alto Research Center
PDA	personal digital assistant
PDR	proliferative diabetic retinopathy
PHC	personal hygiene cost
PMD	personal medical device
PMI	Precision Medicine Initiative
POS	point-of-sale
PR	precision recall
PYLL	Potential Years of Life Lost
QoS	Quality of Service
RFID	radio frequency identification
RNN	recurrent neural network
ROI	region of interest
RPN	region proposal network
SaaS	software as a service
SDN	software-defined networking
SLA	service-level agreement
SOA	service-oriented architecture
SPI	Serial Peripheral Interface
SPP	spatial pyramid pooling
SRA	Sequence Read Archive
SVM	support vector machine
TCP	transmission control protocol
TF-IDF	term frequency inverse document frequency
UART	universal asynchronous receiver/transmitter
UC	ubiquitous computing
UDP	User Datagram Protocol

UPC	universal product code
UX	user experience
VAE	Variational Auto-Encoder
VM	Virtual Machine
VOC	volatile organic compound
VR	virtual reality
YOLO	You Only Look Once

Chapter 1

A New Frontier in Machine Intelligence: Creativity

Ardhendu G. Pathak

1.1 Introduction

On October 25, 2018, the topic of computer creativity, until then mostly an idle curiosity for lay persons and a plaything for computer nerds, suddenly captured the public imagination when a painting called "Portrait of Bellamy" (Figure 1.1) was sold for \$432,000 at Christie's auction house [1]. This painting was created using a type of Artificial Intelligence (AI) program called Generative Adversarial Network (GAN). At the bottom right of the painting, a mathematical equation appeared where normally the artists' signature is placed.

Creativity has long been considered a gift from the gods. For example, the ancient Greeks believed that the Muses – goddesses for different areas of artistic expression, such as epic poetry – whispered in the ears of the mortals to impart the powers of creativity. In India too, the story goes that "*Kavikulguru*" (the Dean of Poets) Kalidas received a boon from the goddess Kali to compose his epic poems in Sanskrit.

The modern view of human creativity, however, considers it a part of the many dimensions of the human intelligence. Being subjective and socially indexed, there is no single accepted definition of creativity. Among many different classification schemas, Boden [2] proposes that creative output can be classified as either combinatorial, exploratory, or transformational creativity. Novel or improbable combinations of familiar ideas are denoted as combinatorial creativity. The exploratory

DOI: 10.1201/9781003424550-1

Figure 1.1   Portrait of Bellamy.

Source: Wikipedia. Artist: Obvious Collective. https://en.wikipedia.org/wiki/ Edmond_de_Belamy#/media/File:Edmond_de_Belamy.png

creativity involves the generation of novel ideas by the exploration of existing ideas or constructs, while the transformational creativity involves generation of novel ideas and constructs by transformation of an existing idea space in such a way that the new constructs bear little or no relation to the existing ones.

The attempts to get computers to do something that can be fun or interesting ("creative") apart from mathematical computations started with the widespread availability of computers. In this chapter, we will first briefly discuss the history of such efforts. We will then see some examples of creative output generated by computers using AI. Next, the legal and intellectual property rights implications of the output autonomously generated by computers will be covered. We will conclude with the outline of the philosophical debate around the artificial creativity domain.

1.2 A Short History of Computer Creativity

As mentioned earlier, people started exploring what computers can do, apart from lightning-fast computations, as soon as computers were available. One such earliest "artwork" in the Victoria and Albert Museum, London, is a long exposure photo of waves on a fluorescent screen of an oscilloscope, obtained by Ben Laposky in 1952 [3].

Another example of "combinatorial creativity" was a program called "Computerized Haiku," written in 1968 by Margaret Masterman and Robin

McKinnon-Wood [4]. It generated short haikus such as the one below, based on the words chosen by the user:

ALL BLUE IN THE HILLS,
I TRACE PALE CLOUDS IN THE DUSK.
HUSH! THE STORM HAS BURST.

A program called JAPE was constructed in 1981 at the University of Edinburgh [5]. It produced short punning puzzles such as:

−What is the difference between leaves and a car?
    One you brush and rake, the other you rush and brake.
−What is the difference between a pretty glove and a silent cat?
    One is a cute mitten; the other is a mute kitten.
−What do you call a strange market?
    A bizarre bazaar.

In the 1960s, for the early computer artists, especially for visual art, one of the significant limitations was the output devices. Plotters, connected with a brush or pen, were used in early experimentations. An example of a complex algorithmic work from that period is a screen-print of a plotter drawing, entitled "Hommage à Paul Klee 13/9/65 Nr. 2," created by Frieder Nake in 1965. The work was inspired by Paul Klee's oil painting entitled "Highroads and Byroads" (1929) [3].

One of the pioneers of computer art or what is now called generative art, was Harold Cohen. He was an accomplished painter and an outstanding engineer. He developed a system called AARON that became increasingly sophisticated (and large in size), and the output was exhibited at prestigious museums and galleries [6]. When asked who was the artist, Cohen or Aaron, Harold gave the example of the relationship between Renaissance painters and their assistants. This question of authorship, which appears quite academic and abstract, has some serious commercial and legal implications. We will discuss this later in the section related to intellectual property (IP) rights.

With the advent of Artificial Intelligence (AI), the "picture" for computer creativity changed dramatically. This was due to the availability of powerful algorithms and computational power coupled with advances in the related fields such as 3D printing and interactive display technologies.

1.3 Artificial Intelligence and Creativity

Applications of AI to creative endeavors have evolved in step with the advances in machine learning and deep learning techniques. The artist + machine collaboration

Figure 1.2   Image generated using VQGAN + CLIP colab notebook. The text prompt was "a cityscape in the style of Van Gogh."

spans a spectrum from an artist as a programmer, with the machine carrying out the instructions, to the learning or evolutionary behavior being so unpredictable as to leave the programmer/artist almost entirely out of the creative process.

The machine learning models used for creative purposes can be classified into two categories: (1) continuous; and (2) discrete or sequential models [7]. The continuous models are used to generate continuous artifacts such as images whereas the sequential models are used mainly to generate artifacts such as text. The models or network topologies used in such application include Variational Auto-Encoders (VAE), autoregressive models, Generative Adversarial Networks (GANs), and sequence prediction models. For more technical details of these models, see [8].

These models have been successfully used to generate images from curated training data sets. Figure 1.2 is a sample image generated by the author using one such program that uses text prompt to generate images and uses a second, "adversarial," network to refine it [9]. The text prompt given in this case was "a cityscape in the style of Van Gogh."

In 2016, an artwork in the style of the Dutch master painter Rembrandt was unveiled by researchers in the Netherlands (Figure 1.3). The artwork, entitled "The

Figure 1.3   The Next Rembrandt.

Source: https://news.microsoft.com/europe/features/next-rembrandt/

Next Rembrandt," was generated using a training set consisting of the known portraits painted by the seventeenth-century master [10].

For Natural Language Processing (NLP) applications, sequence prediction models for text generation have been successfully demonstrated through training sets consisting of either phonemes, syllables, words, or lines. One of the early and fun examples of text generation using recurrent neural networks (RNN) was the Shakespeare generator [11]. A sample below looks quite convincing!

> KING LEAR:
> O, if you were a feeble sight, the courtesy of your law,
> Your sight and several breath, will wear the gods
> With his heads, and my hands are wonder'd at the deeds,
> So drop upon your lordship's head, and your opinion
> Shall be against your honour.

In the field of poetry, Google has an experimental AI-powered tool that can compose poetry snippets, based on the training corpus of the works by classic American poets and another one that can even generate your "poetry portrait" [12, 13]. In 2016, a short Japanese novel written by a program reached the second round of the national literary prize contest [14].

Another interesting example in a creative field was the first AI-scripted movie, whose screenplay was generated by using sci-fi scripts as the training set. The effect, where professional actors enact the AI-generated scenes, is quite hilarious and unexpected. The movie, *Sunspring*, was placed in the top ten at the annual Sci-Fi London Film Festival in one of the categories [15].

Meanwhile in the field of games, AlphaGo, developed by Google DeepMind, defeated Mr. Lee Sedol, the winner of 18 world titles at professional Go, by a 4-1 victory in Seoul, South Korea, in March 2016. The viewership for this game reached 200 million worldwide. During the games AlphaGo played several inventive moves, several of which – including a specific move number 37 in game two – were surprising even for the defending champion. Afterwards Lee Sedol said, "I thought AlphaGo was based on probability calculation and that it was merely a machine. But when I saw this move, I changed my mind. Surely, AlphaGo is creative" [16].

Recently, the company OpenAI announced the Generative Pre-trained Transformer 3 (GPT-) language model [17]. Trained on trillions of words from the internet, it was the largest NLP model till that point. GPT-3 is aimed at natural language answering of questions, but it can also translate between languages and coherently generate improvised text. It has been successfully used for many tasks from copywriting to automatic generation of computer code. Recently, OpenAI also announced a program called DALL·E 2 in the beta testing phase that can create realistic images and art from a natural language description (e.g., "Teddy bears shopping for groceries as a one-line drawing").

Similarly, in the field of music, there are many projects that include Sony Computer Science Laboratories' AI-assisted music production program called Flow Machines. DeepMind, an AI company owned by Google, has created software that can generate music by listening to recordings [18]. At this point we are not aware of a similar project for Indian classical or pop music.

Mathematics is another area where creativity is thought to be a preserve of the select few individuals with natural (or God-given) gifts. Here too AI has been used for discovery and to generate proofs of mathematical theorems. Google has an open source project called DeepHOL that has proved more than 1,200 theorems it had not previously seen [19]. Among the AI-based discoveries are a new formula for pi, and some results in number theory.

1.4 Artificial Creativity: The Debate

The debate about whether something produced by machines can be considered creative had been raging even before the possibility of such machines existed. The discussions have closely followed the similar debate about artificial intelligence [2, 20, 21, 22].

Broadly speaking, the arguments fall into the following groups:

1. Some researchers have proposed a test for artificial creativity on the lines of the Turing Test for judging artificial intelligence as "the philosophy, science and engineering of behaviors that unbiased observers would deem to be creative" [23]. This does not presume any *a priori* criteria and relies only on human judgment.

2. The second camp proposes to define a criterion for creativity and then test to see if any of the machine-generated works would pass muster. For example, Boden [2] defined creativity as "the ability to come up with ideas or artifacts that are new, surprising and valuable." Of course, all three attributes listed here are not objectively measurable and hence necessarily assume human judgment and subjectivity.

3. The third stream of discussion is the one Alan Turing [24] referred to as "Lovelace's objection," after Ada Lovelace, the celebrated English mathematician whom many consider to be the first computer programmer. She wrote: "[The Analytical Engine of Charles Babbage … has no pretensions to originate anything, since it can only do whatever we know how to order it to perform." The famous "Chinese room" thought experiment of John Searle [25] for Artificial Intelligence is in a similar vein. It goes something like this: imagine a scenario where a person ignorant of Chinese language and script assembles sentences according to an instruction manual using cards that have Chinese logograms. The long chain of symbols thus assembled is meaningless to the person assembling it, but to a person who knows Chinese, they may make perfect sense. As per this scenario, the output of any computer program (including AI) might surprise its creators, but it still does not constitute intelligence (or creativity) on the part of machines in any human sense.

As can be seen from the above, the debate is far from settled and there are valid arguments on all sides. But meanwhile, surprising, useful, and sometimes aesthetically pleasing results are being obtained by programs that use increasingly sophisticated techniques and datasets, as seen in Section 1.3.

1.5 Artificial Creativity and Copyright

AI-assisted creation of works could have significant ramifications for copyright law. Traditionally, most definitions of originality to grant the copyright for a work need a human creator, and tools such as a brush or pen don't matter for the purpose of copyright determination. In the same vein, for computer-generated work, the computer and the program used for generation have been seen as mere tools in the creative process. But with the advent of sophisticated AI paradigms, the program is not just a tool; it makes many "creative" decisions to produce an output that is novel, useful and surprising to humans. Therefore, the long-held legal doctrines and policies are being tested in the new AI-driven world [26].

While looking at the copyright and AI-created work, two types of situations have to be considered: (1) how the work "created" by AI should be protected, if at all ("AI as producer"); and (2) to what extent the data and algorithms used in training the AI might constitute fair use and infringement ("AI as consumer") [27].

The question of "authorship" must be decided first when considering the provisions of copyright laws to any work in general, and AI-generated work in particular, since the author is the first owner of the copyright. For AI-generated work, this question is not easy to answer. The candidates for authorship range from the programmer who created the system, the user who might have given final inputs at "runtime," the AI computer system itself, and finally the investors who might have funded the efforts. In most jurisdictions including the US and EU countries, the law does not recognize non-human authorship. In a recent case involving selfies taken by monkeys, the US Court rejected any legal standing for animals to claim copyright [28]. The same logic may then be extended to machine-generated/assisted work.

In terms of "AI as consumer" of the copyright, it is known that lots of data are consumed in training the AI models. Some of this data, whether in terms of text, images, or music, may be owned by third parties. Whether there would be any liability or not resulting from the use of such data would depend on the application of the "fair-use" doctrine that covers transformative and derivative products that are not market substitutes of the original copyrighted material (e.g., making thumbnail images and storing them for training).

The topic of copyright for autonomously generated creative work using machines is quite involved, and the legal thinking is still being shaped in the important jurisdictions [29].

1.6 Conclusion

In this chapter we have seen a brief history of using machines to produce creative work. We also saw the debate this has generated right from the start about the very nature of creativity and whether machines can be creative. In many ways, the arguments, both for and against, parallel those that are familiar from the debate on the nature of artificial intelligence. After outlining representative samples from the early attempts at machine creativity, we saw the virtual explosion of the use of AI in what are traditionally considered creative fields such as painting, writing poetry, proving mathematical theorems, or playing complex strategy games, etc.

The creative output generated using machines autonomously or semi-autonomously also poses some tricky and interesting legal question about ownership and copyright. While the legal frameworks and doctrines are still evolving in this area, the outcomes will have significant effects on multiple industries and research directions in this field.

1.7 Postscript

While this manuscript was in preparation, US-based company OpenAI released their chatbot, ChatGPT, based on a so-called "Large Language Model" (LLM), consisting of billions of parameters. It became a global sensation for its ability to generate text in a variety of writing styles, invent engaging narratives, and even to explain poetry and jokes. LLMs, coupled with the vastly improved image generation programs such as Dall-E and Stable Diffusion, have significantly redefined the landscape of creativity in the context of artificial intelligence and machine learning. As these models continue to evolve, they will reshape our notions of creativity itself. Furthermore, their emergence has sparked a renewed debate on intellectual property rights. Additionally, these models have brought the issues of bias and fairness to the forefront of the conversation, as their responses are influenced by the data they are trained on. The implications of these developments extend far beyond the academic sphere, influencing industries ranging from literature to advertising.

References

1. Christie's (n.d.) "Is Artificial Intelligence Set to Become Art's Next Medium?" Available at: www.christies.com/features/A-collaboration-between-two-artists-one-human-one-a-machine-9332-1.aspx (accessed March 25, 2022).
2. M. A. Boden (1998) "Creativity and Artificial Intelligence." *Artificial Intelligence*, 103(1): 347–356, doi:10.1016/S0004-3702(98)00055-1.
3. Victoria and Albert Museum (2011) "The Pioneers (1950–1970)." Available at: www.vam.ac.uk/content/articles/a/computer-art-history/ (accessed April 19, 2022).
4. "Computerized Haiku." Available at: www.in-vacua.com/cgi-bin/haiku.pl (accessed April 20, 2022).
5. G. Ritchie (2003) "The JAPE Riddle Generator: Technical Specification." Institute for Communicating and Collaborative Systems, January (accessed April 18, 2022). Online. Available at: www.academia.edu/403908/The_JAPE_Riddle_Generator_Technical_Specification
6. CHM (2016.) "Harold Cohen and AARON—A 40-Year Collaboration." Blog. August 23. Available at: https://computerhistory.org/blog/harold-cohen-and-aaron-a-40-year-collaboration/ (accessed March 25, 2022).

7. G. Franceschelli and M. Musolesi (2021) "Creativity and Machine Learning: A Survey." *arXiv*, April. Online. Available at: http://arxiv.org/abs/2104.02726 (accessed March 24, 2022).

8. S. J. Russell, S. Russell, and P. Norvig (2020) *Artificial Intelligence: A Modern Approach.* Pearson, Online. Available at: https://books.google.co.in/books?id=koFptAEACAAJ

9. "Google Colaboratory." Available at: https://colab.research.google.com/drive/1NCceX 2mbiKOSlAd_o7IU7nA9UskKN5WR?usp=sharing#scrollTo=NRUouuHptoRp (accessed May 5, 2022).

10. Microsoft News Centre Europe (2016) "The Next Rembrandt: Recreating the Work of a Master with AI." Available at: https://news.microsoft.com/europe/features/next-rembra ndt/ (accessed April 28, 2022).

11. "The Unreasonable Effectiveness of Recurrent Neural Networks." Available at: http:// karpathy.github.io/2015/05/21/rnn-effectiveness/ (accessed April 25, 2022).

12. "POEMPORTRAITS." Available at: https://artsexperiments.withgoogle.com/poempo rtraits (accessed May 6, 2022).

13. "Verse by Verse." Available at: https://sites.research.google/versebyverse/ (accessed April 28, 2022).

14. Digital Trends (2016) "Japanese A.I. Writes Novel, Passes First Round for Literary Prize." Available at: www.digitaltrends.com/cool-tech/japanese-ai-writes-novel-passes-first-round-nationanl-literary-prize/ (accessed March 25, 2022).

15. A. Newitz (2021) "Movie Written by Algorithm Turns Out to Be Hilarious and Intense." *Ars Technica*, May 30. Available at: https://arstechnica.com/gaming/2021/05/an-ai-wrote-this-movie-and-its-strangely-moving/ (accessed April 25, 2022).

16. DeepMind "AlphaGo: The Story So Far." Available at: https://deepmind.com/research/ case-studies/alphago-the-story-so-far#our_approach (accessed March 24, 2022).

17. "Better Language Models and Their Implications." Blog. Online. Available at: https:// openai.com/blog/better-language-models/ (accessed March 25, 2022).

18. Magenta. "Magenta." Available at: https://magenta.tensorflow.org/ (accessed May 6, 2022).

19. Math Scholar (2019) "Google AI System Proves Over 1200 Mathematical Theorems." Available at: https://mathscholar.org/2019/04/google-ai-system-proves-over-1200-mathematical-theorems/ (accessed April 27, 2022).

20. *MIT Technology Review* (2019) "A Philosopher Argues that an AI Can't Be an Artist." Available at: www.technologyreview.com/2019/02/21/239489/a-philosopher-argues-that-an-ai-can-never-be-an-artist/ (accessed March 24, 2022).

21. A. Mello (2020) "Creativity and Artificial Intelligence." *Medium*, May 22. Available at: https://towardsdatascience.com/creativity-and-artificial-intelligence-46de4326970c (accessed March. 24, 2022).

22. G. Franceschelli and M. Musolesi (2022) "Deep Creativity: Measuring Creativity with Deep Learning Techniques." *arXiv*, January. Online. Available at: http://arxiv.org/abs/ 2201.06118 (accessed March 24, 2022).

23. F. Carnovalini and A. Rodà (2020) "Computational Creativity and Music Generation Systems: An Introduction to the State of the Art." *Frontiers in Artificial Intelligence*, 3: 14. doi:10.3389/frai.2020.00014.

24. A. M. Turing (1950) "I.—Computing Machinery and Intelligence," *Mind*, LIX(236): 433–460., doi:10.1093/mind/LIX.236.433.

25. J. R. Searle (1980) "Minds, Brains, and Programs," *Behavioral and Brain Sciences*, 3: 417–424.
26. WIPO (2017) "Artificial Intelligence and Copyright." Available at: www.wipo.int/wipo_magazine/en/2017/05/article_0003.html (accessed March 24, 2022).
27. E. Bonadio and L. McDonagh (2020) "Artificial Intelligence as Producer and Consumer of Copyright Works: Evaluating the Consequences of Algorithmic Creativity," *Intellectual Property Quarterly*, 2: 112–137. Online. Available at: www.sweetandmaxwell.co.uk/Catalogue/ProductDetails.aspx?recordid=380&productid=6791 (accessed March 24, 2022).
28. Wikipedia (2022) "Monkey Selfie Copyright Dispute." May 1. Online. Available at: https://en.wikipedia.org/w/index.php?title=Monkey_selfie_copyright_dispute&oldid=1085659976 (accessed May 3, 2022).
29. J. K. Eshraghian (2020) "Human Ownership of Artificial Creativity." *Nature Machine Intelligence*, 2(3). doi:10.1038/s42256-020-0161-x.

Chapter 2

Overview of Human-Computer Interaction

Nancy Jasmine Goldena

2.1 Introduction

HCI definition: Human-computer interaction (HCI), also known as man-machine interaction (MMI) or computer-human interaction (CHI), is the study of interactive computing systems for human use, including their design, evaluation, and implementation.

HCI defined by experts: Stuart K. Card, Allen Newell, and Thomas P. Moran popularized the term HCI in their book, *The Psychology of Human Computer Interaction*, in 1983. Carlisle was the first to coin it in 1975. Experts have defined HCI as "computers, unlike other technologies with specialized and limited functions, offer a wide range of applications that typically necessitate an open-ended interaction between the user and the computer" [1].

HCI is important as it will be necessary for products to be more successful, safe, beneficial, and functional. It will make the user's experience better in the long run. As a result, having someone with HCI skills involved in all phases of any product or system development is critical. HCI is also necessary to prevent products or projects from going wrong or failing completely.

HCI aims to develop truly effective and useful interfaces and screens with today's modern technology and resources. Its main purpose is to improve interactions by making computers more accessible and adaptable to the user's needs and the long-term goal is to build systems that reduce the wide gap between the human's cognitive perspective about

DOI: 10.1201/9781003424550-2

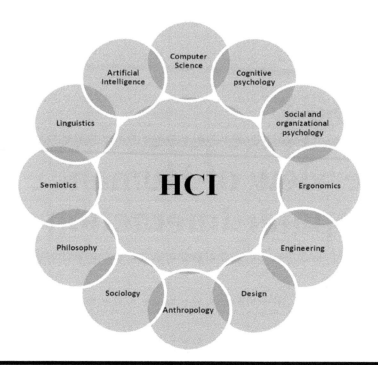

Figure 2.1 Human-computer interaction.

what they wish to do and the computer's ability to comprehend such a user's activity. HCI can be implemented in any sector where computer installation is possible.

In general, HCI is the study of how to create computer systems that efficiently meet people's needs. HCI incorporates information and methodologies from a variety of fields [2], including computer science, cognitive psychology, social sciences, and ergonomics (human factors) (Figure 2.1).

Computer science includes technological knowledge as well as developing abstractions, methodologies, languages, and tools to handle the problem of creating a successful design with inspiration and an understanding of what people want. Cognitive psychology is a branch of psychology that analyses mental processes, such as perception, memory, problem-solving, emotions, and learning in order to gain insight into users' capabilities and limitations. It focuses on users and their tasks rather than the technology they use. Ethnomethodology is a technique used in social psychology to understand the structure and functions of organizations and addresses human behavior with respect to context, society, and interactions. Ergonomics (also known as human factors) is a scientific discipline that studies the interactions between humans and other system elements, as well as a profession that uses theory, concepts, data, and methodologies in order to improve human well-being and overall system performance. Sociology is a social science that studies society's influence on human behaviors. It is particularly concerned in how societal pressures influence human behavior. Individual behavior is examined in psychology, and it offers a scientific viewpoint on how individual strengths and limitations can be considered when designing effective systems. By providing a framework for navigating, experiencing, understanding, and assessing the world and its objects, philosophy offers an approach to how human abilities and limitations might be taken into account in the design of effective systems.

Artificial Intelligence (AI) is inspired by human intelligence, made powerful by human data, and ultimately beneficial in how it improves the human experience and empowers researchers to make better decisions. Anthropology is the study of humanity, and it helps to improve usability and user-centrism by involving users in the design process. It specifically aims to understand future users of a design, with the goal of allowing the design team to develop an empathetic understanding of the users' practices, routines, and values. Linguistics is the study of human language as a science. It comprises a thorough, methodical, objective, and precise examination of all aspects of language, particularly its nature and structure, and it aids in the generation of demographic stereotypes of physical-to-abstract mappings as well as the inspiration for user interface design. Semiotics is crucial for designers because it enables them to comprehend the links between signs, what they represent, and the people who must interpret them as well as the people for whom the design is created.

2.2 The Aim of HCI

The aim of HCI is to create systems that are both useful and safe, as well as functional. To create usable computer systems, developers must strive to do the following:

- improve user-computer interactions;
- understand the factors that influence how people use technology;
- develop tools and techniques to enable the creation of appropriate systems;
- achieve efficient, effective, and safe interactions;
- put people first.

2.3 Factors in HCI

In order to analyse and build a system employing HCI principles, a significant number of aspects must be examined. Many of these variables interact, thus complicating the investigation. The following are the most important factors affecting HCI [3].

1. *Organization factors*: training, job design, roles, politics, work organization
2. *Environmental factors*: noise, humidity, heating, lighting, ventilation

3. *Health and safety factors*: stress, headaches, musculo-skeletal disorders
4. *User factors*: cognitive process, capabilities, motivation, enjoyment, satisfaction, personality, level of experience
5. *Comfort factors*: seating, equipment, layout
6. *User interface*: input devices, output devices, dialogue structures, colors, icons, commands, navigation, graphics, natural language, 3-D, user support materials, multi-media
7. *Task factors*: easy, complex, novel, task allocation, repetitive, monitoring, skills, components
8. *Constraints*: costs, timescales, budgets, staff, equipment, building structure
9. *System functionality factors*: hardware, software, application
10. *Productivity factors*: increase output, increase quality, decrease costs, decrease errors, decrease labor requirements, decrease production time, increase creative and innovative ideas leading to new products.

2.4 HCI Design Issues

HCI design should be user-centered and involve users as much as possible so that they can influence it; it should also integrate knowledge and expertise from the various disciplines that contribute to HCI design, It should also be highly iterative so that testing can ensure that the design meets the users' requirements.

There are various challenges in HCI design, as follows [4]:

1. Human-machine symbiosis
2. Human-environment interactions
3. Ethics, privacy and security
4. Well-being, health and eudaimonia
5. Accessibility and universal access
6. Learning and creativity
7. Social organization and democracy.

2.4.1 Human-Machine Symbiosis

1. *Meaningful human control*: It is necessary to address the issue of design trade-offs between human control and automation. Machine learning systems must demonstrate their ability to explain and convey knowledge in a way that humans can comprehend. The system's operation must be well understood by the users and the users need to have a deeper understanding of the underlying computational processes in order to better regulate the system's actions.
2. *Human digital intelligence*: A reliable technology needs to be established that supports and respects individual and social life, respects human rights and

privacy, supports users in their activities, creates trust, and allows humans to exploit their individual, creative, social, and economic potential, as well as living and enjoying a self-determined life.

3. *Adaptation and personalization to human needs*: Other organizational and societal aspects, such as the culture of the organization and the user's level of expertise in the use of technology to acquire Big Data, should be considered in personalization/ adaptation, in addition to individuals' preferences.

4. *Human skills support*: Instead of computers that objectify knowledge, a method that requires instruments to assist human talent and inventiveness is required. AI needs to be integrated to support human memory as well as human problem-solving, particularly in instances when a user must handle a large quantity of sophisticated information rapidly or under stress. Human memory and learning capacity can be increased by ensuring that people have access to networked cognitive intelligent technology that can help them solve common difficulties. Through the convergence of ICT and the biological brain, technology is employed to increase human perception and cognition.

5. *Emotion detection and simulation*: Technology must display emotions and empathetic behavior in order to record and correlate the manifestation of human emotions. Deception, whether intentional or accidental, must be dealt with.

   a. *Human safety*: Advanced intelligent systems that are "safe by design" are recommended. New testing approaches must be implemented that not only test the design and implementation of the system, but also the system's gained knowledge. Through focused investigations, monitoring, and analysis, an interdisciplinary group of specialists will provide evaluations and suggestions for a safety system.

   b. *Cultural shift*: Intelligent surroundings and AI provide a cultural transformation in addition to technological hurdles and adjustments. As a result, technological advances must progress in tandem with societal changes. To persuade the public and the scientific community, extensive studies of the ethical and privacy concerns, as well as design methods with their underlying perspectives on the role of people, are required.

2.4.2 Human-Environment Interactions

1. *Interactions in the physical and digital continuum*: The computer no longer appears as a recognizable object. For vanishing computers and devices, new kinds of affordances must be created. It is necessary to frame user commands successfully for the right interactive artifacts. The right user command interpretation should be delivered based on the present context. Recognizable interaction is required for everyday objects, and also information overload might be possible.

2. *Implicit interactions*: It is crucial to consider how to support the constant changes in the interplay between the center and the periphery of attention. In any case, the interactive environment should not place heavy demands on users' perception and cognition or lead to other unfavorable outcomes, such as confusion or dissatisfaction. Implicit interactions in intelligent settings raise privacy issues in addition to design issues.

3. *Novel and escalated interactions*: To distinguish emotions, signify explicit inputs, and identify multimodal communication behavior, more natural multisensory contact should be established, such as taste and smell as well as haptic sensations. In order to meet the additional problems in gathering and structuring the prospective contexts of use, eliciting and evaluating user needs, creating designs, and conducting evaluations, existing approaches need to be scaled up, which calls for new design methodologies.

4. *Interactions in public spaces*: How can interactive systems catch people's attention and encourage them to participate with the system? The prevalence of technology in public spaces that "blur" the lines between private and public contact is another issue for interface design. New techniques for user involvement in the design process are also necessary. The balance between single-user and multi-user situations, as well as the promotion of cooperation between several users who may be strangers, are additional difficulties resulting from the requirement to service multiple users at once.

5. *Interactions in virtual and augmented reality (AR)*: The aim is to create hardware that provides realistic experiences that exhibit greater feelings of presence and immersion. To achieve this, one must enhance the feeling of embodiment, which encompasses the sense of self-location, sense of action, and sense of ownership over the body. As deep and realistic virtual reality (VR) moves closer to true virtuality, there is a serious risk that users may become overly dependent on the virtual environments and virtual characters. Interconnected VR design principles must be improved and further developed to take into account social interactions in virtual spaces. How AR will blend the real and virtual in a way that prevents the experience from being taken solely from either the real or the virtual content is a significant problem.

6. *Evaluation*: Among the difficulties are the interpretation of signals from various communication channels in the context of natural interaction, context awareness, the inappropriateness of task-specific measures in systems that frequently lack tasks, and the requirement for long-term studies to evaluate users' learning processes. In order to provide holistic and systematic ways for the evaluation of user experience (UX) in intelligent settings, taking into account a wide variety of traits and characteristics of such environments, new frameworks and models are required. For the development of AI applications, services, and artefacts, clear standards and a code of ethics should be established.

2.4.3 Ethics, Privacy, and Security

1. *Fundamental privacy concerns*: The numerous privacy problems to be solved include the right to be alone without interruption (solitude), the right to have no public personal identity (anonymity), and the right not to be monitored (intimacy). The right to control one's personal information must be protected as well.

2. *HCI research*: The use of vulnerable user populations in HCI research, such as older adults, people with disabilities, immigrants, socially isolated individuals, patients, children, and so on, should be avoided. Participant anonymity is also a difficult task, as studies have shown that data can be de-anonymized when combined with other datasets.

3. *Online social networks (OSNs)*: Concerns about privacy in OSNs include data retention issues, the ability to browse private information, data selling, and targeted marketing. Privacy and content sharing are two opposing fundamental features of OSNs that must be perfectly combined, a challenge that OSN designers must master.

4. *Healthcare technologies*: In the context of healthcare, data privacy, accuracy, integrity, and confidentiality are essential. Access to eHealth for all, anonymity, autonomy, beneficence and non-maleficence, dignity, no discrimination, free and fully informed consent, justice, safety, and value-sensitive design are additional ethical principles that should be addressed in eHealth.

5. *Virtual reality*: In a virtual reality environment, the illusion causes the user's reactions and feelings, such as over-attachment to virtual agents or feeling out of control and behaving aggressively in both the virtual and physical worlds. The intentions of the creator of the VR environment may be ambiguous and dangerous to the user. Users' perceptions of reality may be affected as a result of VR, resulting in reality dissonance.

6. *IoT and Big Data*: Data about people in the virtual world are supplemented and integrated with real-world data, and data privacy is further jeopardized by the interconnectedness of common objects. The connection of daily items threatens data anonymity even more. By connecting multiple systems and merging data, risks to privacy and identity are realized. Discrimination and restriction of options may result from automated decision-making. Malevolent attempts and deceptive promises should not be made to vulnerable people.

7. *Intelligent environments*: Intelligent systems, in general, pose a number of risks, such as user identification based on collected data, the persistence of personal/sensitive data, profiling and implicit deduction and attribution of new properties to individuals, data use for monitoring, data misinterpretation, public disclosure of confidential information, and data collection and persuasion techniques used without the user's knowledge. Transparency is essential when autonomous intelligent agents make more complicated and significant ethical judgments, so that humans can understand, predict, and appropriately trust AI, whether manifested as traceability, verifiability, non-deception, or intelligibility.

8. *Cybersecurity*: Because of the large number of networked devices, typical security solutions are rendered ineffective. Malicious assaults on sensors and actuators might undermine physical security and trust. The greatest vulnerability in cybersecurity is the human agent.

2.4.4 Well-Being, Health, and Eudaimonia

1. *Personal medical devices (PMDs) and self-tracking*: PMDs have great potential and should be accepted by the medical community. Data controllability, data integration and accuracy, data visualization, input complexity, privacy, aesthetics, and engagement have been identified as barriers to self-tracking adoption. The digitization and automation of self-tracking raise privacy problems as well as ethical considerations about the use of the data gathered. Developing self-tracking technology is much more than a technical endeavor; hence, a multidisciplinary strategy involving software developers, interface designers, doctors, and behavioral scientists should be used.

2. *Serious games for health*: One significant problem is that production of games for health demands high design complexity and multi-disciplinary teams, preferably including users in the process, both of which result in slower implementation. Despite high costs and an underdeveloped industry, consumer expectations are high. Another concern is the assessment of serious games with target consumers.

3. *Ambient assisted living*: Person-centered care must be prioritized through enhancing customization of treatment and taking into account people's diverse requirements, expectations, and preferences. Privacy and confidentiality should be preserved, and technology should not be used to replace human care, resulting in the isolation of older people, or abuse or violation of human rights.

4. *Intelligence in healthcare*: Over-reliance on technology may undermine people's ability to manage their lives and result in temporary loss of abilities. Over-dependence can also result in patient isolation and a lack of personal care. For patient-centric design, new design and assessment approaches are required. To ensure the suitability of technologies, new development and testing techniques are required.

5. *Well-being and eudaimonia*: Difficulty in identifying and quantifying happiness and eudaimonia. The broad scope of these concepts extends beyond personal well-being to community well-being and planet goodness.

2.4.5 Accessibility and Universal Access

1. *Adoption of proactive approaches*: Proactive methods do not support a "one-size-fits-all" approach; rather, they seek to enhance accessibility for all by

tailoring the design to each user. As a result, many businesses regard universal design as an additional cost or feature, which is not the case.

2. *Population aging*: A major concern is how senior users will be encouraged to adopt ICT technologies, as well as how acceptable they will find them, both of which are predicted to alter in the near future as a new generation of technologically-adept elder users emerges.

3. *Accessibility in technologically enriched environments*: The variety of gadgets, programs, and environment intelligence can put a lot of cognitive strain on consumers, if these programs are not properly built. Intelligent settings must be transparent to their users, but this in itself presents a serious accessibility issue, especially for the elderly and people who suffer from cognitive disabilities.

4. *Methods, techniques, and tools*: New evaluation methodologies and tools must be targeted in order to take into account the changing human needs and context of use, develop appropriate user models, advance knowledge of user requirements, and determine whether various solutions are appropriate for different combinations of user and environment characteristics.

5. *Universal access in future technological environments*: In order to prevent anyone from being excluded, isolated, or exploited by the new futures, special consideration must be given to any person who is at risk of exclusion. Age, education, occupational status, health, social connection, infrastructure accessibility, impact on digital inequality, including race and ethnicity, gender, and socio-economic level are some of the key elements that must also be taken into account.

2.4.6 Learning and Creativity

1. *New generation of learners*: It is vital to understand the influences of human factors in order to design digital learning environments that best suit each and every learner by keeping the human at the center and studying the various needs and requirements of each individual.

2. *Extended reality*: A difficult aspect of the design of extended reality technologies is how to use them to support the entire educational process, from the creation of the content to the delivery of the learning experience. This requires a multidisciplinary approach and the active involvement of tutors, who must use these technologies themselves and act as knowledge-conveyors in addition to content providers.

3. *Mobile learning*: The use of mobile devices across the traditional boundaries between formal and informal contexts, virtual and physical worlds, and planned and emergent spaces raises issues about how learning activities should be designed by instructors and how educators' and learners' thinking is reconceptualized.

4. *Serious games for learning*: Designing for and measuring fun in the user experience of serious games is a crucial issue. It is important to strike the correct balance between seriousness and gamification: if learners are exposed to too much educational information, their motivation may decline; conversely, if too much fun is included, learning may suffer. To ensure that game objectives and learning objectives are aligned, all stakeholders need to participate in and collaborate on design processes, metrics, and assessment methods.

5. *Intelligent environments*: How to give learners the appropriate material at the appropriate time and in the appropriate manner presents a significant difficulty, especially when many users are present in such settings and judgments made by one user may be influenced by those of other users. The difficulty lies in finding ways to increase human learning capacity while attempting to make learning efficient and simple.

6. *Pedagogical impact of learning technologies*: A significant issue is that technology should put an emphasis on the viewpoints and requirements of the learners themselves and push the frontiers of education in order to include factors besides grades, such as motivation, self-assurance, fun, contentment, and achieving professional goals. The fact that the same technology may perform differently in various economic, social, and cultural contexts presents another barrier to the effective pedagogical use of technology in education. As a result, a "one-size-fits-all" strategy would not be practical and technology should be customizable and adaptable.

7. *Creativity*: The comprehensive support of the entire creative process, the automatic retrieval and dynamic delivery of stimuli for varying degrees of relevance to the creation task, and the provision of individualized assistance for each participant in a collaborative creation process are indicative developments. Tools will also need to re-evaluate how to encourage creativity in physical environments when the digital and physical are combined.

2.4.7 Social Organization and Democracy

1. *Sustainability*: The practice of systems thinking is one that is gaining popularity. Stakeholders and advanced device ecologies must be involved in the design of technology. The design of technologies must recognize a future in which resources are scarce, in order to handle issues related to crisis response, and face conditions in which infrastructure availability may be low, healthcare provision may be inadequate, food supply may be unstable, and governments may be ineffective or corrupt.

2. *Social justice*: Who benefits from a technology and whether it can be designed in a more inclusive way to benefit other socio-economic classes more equally should be the main concerns when it is being developed from a larger social justice perspective.

3. *Active citizen participation*: Participation of citizens in the design and scientific endeavors creates a number of issues relating to user participation, processes, and outcomes, as well as the technology itself. How to engage citizens in meaningful involvement and how to promote ongoing engagement and inclusiveness are two pressing participation-related issues.

4. *Democracy*: Fake news, echo chambers, and agenda-setting through enhanced awareness of the most popular articles in the media are indications of technological threats to democracy. Monopolies in social media technology pose the risk of reshaping humanity into what they want it to be. These worries are made worse by advanced Big Data and AI technologies, which have the potential to create an automated society with authoritarian elements in which AI would have complete control over our knowledge, opinions, and behavior.

These are the top seven technological and sociological issues that need to be addressed in order to respond to urgent societal and human requirements when using the ever-increasing interaction intelligence.

2.5 HCI Implementation Issues

User interfaces are not only difficult to design, but they are also difficult to implement. The problems associated with user interfaces are as follows [5]:

1. Need for iterative design
2. Reactive programming
3. Multiprocessing
4. Need for real-time programming
5. Need for robustness
6. Low testability
7. No language support
8. Complexity of the tools
9. Difficulty of modularization.

We will now discuss these issues in detail:

1. *Need for iterative design*: The traditional "waterfall" approach to software design, in which the user interface is fully designed, then implemented, and finally tested, is insufficient. Rather, specification, implementation, and testing must all be integrated. This makes scheduling and managing user interface development extremely challenging.

2. *Reactive programming*: User interface software is more complex than other types of software. One significant difference is that current user interfaces

must be developed from the inside out. Rather than organizing the code so that the application has control, the program must be designed as a series of sub-routines that the user interface toolkit calls when the user does something. Each sub-routine will have strict time limits so that it completes before the user is ready to issue the next command, making it more difficult to organize and modularize reactive programs.

3. *Multiprocessing*: User interface software is typically structured into various processes in order to be reactive. As a result, programmers developing user interface software will encounter well-known issues with numerous processes, such as synchronization, thread consistency, deadlocks, and race situations.

4. *Need for real-time programming*: Objects that are animated or move around with the mouse are common in graphical, direct manipulation interfaces. To be appealing to users, the objects must be presented between 30 and 60 times per second with no unequal pauses. As a result, the programmer must ensure that any processing required to calculate the feedback may be completed in less than 16 milliseconds. This may include employing less realistic but faster approximations and complex incremental algorithms that compute the output based on a single changing input rather than a simpler recalculation based on all inputs.

5. *Need for robustness*: The program that handles user input has particularly rigorous requirements since all inputs must be treated graciously. A programmer must define the interface to an internal method to only work when a specific type of data is supplied, while the user interface must always take any conceivable input and continue to operate. Furthermore, unlike internal procedures, which may abort to a debugger when an incorrect input is detected, user interface software must respond with a helpful error message and allow the user to restart or repair the error and continue. User interfaces should also allow the user to cancel and undo any operation. As a result, the programmer must design all activities so that they can be canceled during execution and reversed after completion. To accommodate this, special data structures and coding styles are usually required.

6. *Low testability*: Because automated testing methods have problems supplying input and testing output for direct manipulation systems, user interface software is more challenging to test.

7. *No language support*: Programming user interface software is challenging since today's programming languages lack necessary features such as primitives for graphical input and output, as well as reactive programming and multi-processing.

8. *Complexity of the tools*: A huge number of tools designed to improve the software's user interface are notoriously difficult to use. The tool manuals are often many volumes long and contain hundreds of procedures, therefore

learning to program user interfaces with these tools often requires extensive training. Despite their size and complexity, the tools may not even provide enough flexibility to accomplish the intended result.

9. *Difficulty of modularization*: One of the main ways to make software easier to design and maintain is to properly modularize the various elements. The user interface should be segregated from the rest of the software, in part so that it may be quickly modified (for iterative design). Unfortunately, programmers discover in practice that it is difficult or impossible to segregate the user interface and application elements, and modifications to the user interface frequently necessitate reprogramming parts of the application as well.

While designing and implementing complex software are challenging in general, user interfaces appear to bring considerable additional obstacles, and we may expect research into user interface design and implementation to continue to deliver better theories, approaches, and tools in future.

2.6 Important Aspects of HCI

The assurance of user satisfaction is a key component of HCI. Because HCI examines how humans and machines interact, it draws on relevant knowledge from both the human and the machine sides. The summary of key HCI aspects and definitions are as follows [6]:

- *Effectiveness*: The precision and thoroughness with which users accomplish their intended objectives.
- *Efficiency*: The amount of resources used in comparison to how accurately and completely users fulfill their goals.
- *Satisfaction*: A good attitude toward using the system's environment and freedom from discomfort.
- *Flexibility*: The extent to which a system can accommodate changes desired by the user beyond those first specified.
- *Learnability*: The amount of time and effort needed to master a given level of system performance, also known as the learning curve.
- *Memory*: The time and work needed to get back to a certain level of performance after being absent from the system for a certain amount of time.
- *Safety*: The components of a system that are concerned with defending the user against risky situations and unwanted situations.
- *Usability*: The degree to which a system can be employed by specific users to accomplish particular goals in a specific context with effectiveness, efficiency, and satisfaction. Usability is the general technical term for user-friendliness. Usability also includes adaptability, memory, learnability, and safety.

- *Accessibility*: A system's environment that is most usable and has the most capabilities. Web accessibility refers to the ability of persons with impairments to interact, perceive, comprehend, and navigate the web.
- *User experience*: Beyond efficacy, efficiency, and satisfaction, this covers the elements of user interactions with and responses to systems.

Designers must keep in mind and examine these essential aspects and issues when creating and implementing HCI systems.

HCI is a scientific and practical sector that first developed in the late 1970s and early 1980s as a sub-field of computer science. HCI has grown rapidly and consistently, attracting individuals from a variety of disciplines and encompassing a wide range of concepts and methodologies.

During the 1980s and 1990s, HCI was primarily concerned with developing systems that were simple to understand and use. An individual could not afford a computer at that time. Computers developed into communication tools between the 1990s and the early 2000s. It became essential to examine external impacts and how interactions develop across technologies and enterprises. During this time, email became ubiquitous, which meant that people were not only connecting with computers, but also with one another through computer systems.

During the 2000s and 2010s, value-driven innovation took the lead in implementing initiatives and designing interfaces for long-term development. A new interface design approach is needed for complicated connections among people, environments, and tools [7].

Currently, Ubiquitous Computing (UC) is now the most active research field in HCI. UC and immersive technologies such as augmented reality (AR) are becoming extremely important in HCI.

2.7 Components of HCI

A computer interface is a medium that allows any user to communicate with a computer. HCI is made up of three main components:

- a human
- the computer
- the interaction between them.

The interface is essential to most HCI. To create effective interfaces, users must understand the limitations and capabilities of the components. To accomplish various tasks, humans interact with computer systems. Input-output channels vary between humans and computers [8]. Input and output channels for human and computer components are briefly discussed here.

Human components are classified as follows:

- *Long-term memory*: This has the ability to store data for a long time.
- *Short-term memory*: All information that enters the system is stored in short-term memory, for a limited time.
- *Sensory memory*: Sensory memories are short-term memories that occur in our sensory organs.
- *Visual perception*: Visual perception is the ability to perceive the environment through the eyes, including colors, patterns, and structures.
- *Auditory perception*: Auditory perception is the ability to receive and comprehend information provided to the ears via audible frequency waves transmitted through the air.
- *Speech and voice*: Speech has become a more frequent mode of communication with computer systems. The most popular examples are Amazon Echo, Apple's Siri, and Google Home.

Computer components are classified as follows:

- *Text input devices*: A text input device, sometimes known as a text entry device, is an interface for entering text data into an electronic device. A mechanical computer keyboard is a well-known example of an input device.
- *Speech recognition*: This is a feature that allows a computer to transform human speech into text, also known as automatic speech recognition (ASR).
- *Mouse/touchpad/keyboard*: A mouse is a small, movable device that can be wired or wireless and allows the user to control a variety of functions on a computer. Used with the keyboard to control input.
- *Eye-tracking*: Eye-tracking is a sensor technology that can detect and monitor the movement of a person's eyes in real time.
- *Display screens*: The graphic design and layout of user interfaces on displays are referred to as screen design.
- *Auditory displays*: The use of audio to convey information from a computer to a user is known as auditory display.
- *Printing abilities*: Users have the option of printing their own customized user interface.

2.8 The Characteristics of HCI

HCI can be applied to a wide range of situations. Some of the essential characteristics of HCI are as follows:

- *Shapes and assembles objects to change situations*: Interaction designers create objects to modify the way humans interact with technology. When a user is in a changing scenario, a different type of interface design may be required. Interaction design should be focused on digital design in particular.

■ *Investigates future scenarios*: Interaction designers require flexibility to investigate and innovate, which involves exploring the future in a variety of ways before choosing a design method.

■ *Parallels to the problem with alternative solutions*: A good design with many alternative solutions is required. Sometimes an amazing design does not have a great solution.

■ *Necessary to understand in terms of "sketching" and other "physiological representations"*: Interaction designers create tangible (can be touched) models to test ideas and concepts to observe if users like or dislike them. Sketches are the most basic technique to model a core idea. However, complete and comprehensive prototypes may be required to truly understand how a user interacts with a particular prototype.

■ *HCI considers instrumental, technical, artistic, and ethical issues throughout the process*: Innovations must be functional, attractive, technologically possible and acceptable [9].

2.9 HCI Principles and Best Practices

Human perception and information processing principles can be used to build a successful display. Certain principles may or may not be applicable to various scenarios. Some principles may appear to be contradictory and there is no clear answer to determine which is more significant. These rules range from basic design ideas to best practice [8]. This section provides important HCI design rules that are generally followed.

■ Learnability and familiarity determine how quickly a novice user can learn to interact with a system. To do so, a system must be predictable.

■ To be effective, all characters and objects must be available.

■ Users will make mistakes, but the mistakes should not interrupt the system or have a negative impact on an interaction. To prevent potential mistakes, warnings should be provided and the system should also offer assistance when potential errors occur.

■ Any non-trivial critical action should be authenticated.

■ Reduce the amount of information that needs to be recalled in the gaps between tasks.

■ The system may struggle to meet the user's needs if the information given is insufficient or inconsistent.

■ Don't overload the user with data; instead, choose a presentation layout that enables an efficient information system.

■ Use common labels, abbreviations, and colors that are most likely.

- Allow the user to keep their visual context.
- Create informative error messages.
- To improve comprehension, use upper and lower case, indentation, and text grouping.
- To categorize different kinds of data, use frames.
- Use analog displays to represent information that is easier to fit.

Several HCI principles have been suggested by three experts: Ben Shneiderman, Donald Norman, and Jakob Nielsen.

The following eight broad guidelines were developed by Ben Shneiderman, an American computer scientist, who documented certain hidden truths regarding design:

1. Consistency is important.
2. Universal usability should be considered.
3. Provide constructive feedback.
4. Create dialogs to achieve closure.
5. Errors should be avoided.
6. Allow easy action reversal.
7. Encourage internal locus of control.
8. Short-term memory load should be reduced.

Using these eight rules, you can tell the difference between a good and a bad interface design. These are useful in the experimental evaluation of better graphical user interfaces (GUIs).

Donald Norman recommended seven stages for transforming challenging tasks:

1. Use both real-world and intellectual knowledge.
2. Reduce task complexity.
3. Make things clear.
4. Make sure your mapping is correct.
5. Convert limitations into benefits.
6. Plan for mistakes.
7. Standardize when everything fails.

These principles can be used to perform difficult tasks.

Heuristics evaluation is a scientific process for detecting usability issues in user interfaces. The heuristic evaluation approach incorporates various usability concepts, such as Jakob Nielsen's ten usability principles:

1. System status visibility.
2. Relationship between the system and the real world.

3. User control and independence.
4. Standards and consistency.
5. Error minimization.
6. Rather than recall, recognize.
7. Usefulness and adaptability.
8. Minimal and attractive design.
9. Error assistance, diagnosis, and recovery.
10. Documentation and support.

The ten Nielsen principles listed above serve as a checklist for the heuristic evaluator while analyzing a design [8, 10].

2.10 Design for HCI

HCI design is a problem-solving process that includes features such as intended use, target area, resources, cost and viability. Goal-driven design is one of the most important aspects of interaction design. The product design should be simple to operate. Because users don't remember all the functionalities after using a product, the user needs to build familiarity into every interface. To reduce complexity, consistency and predictability must be established [8]. Table 2.1 presents the three stages of product design.

2.10.1 HCI Design Approaches

HCI is essential in building intuitive interfaces that people of various skills and expertise can use. Most notably, human-computer interaction is beneficial to individuals who lack information and formal training on how to connect with specific computing systems.

Table 2.1   Steps in the General Design Process

Steps	Description
Step 1	Before designing an interface, system designer needs to understand the user's requirements and the difficulties experienced by the user before designing an interface. This is known as user research and analysis.
Step 2	The second step is developing and prototyping the designs. In essence, transforming user's requirements into possible solutions. Both conceptual and physical design will be incorporated. Then the prototype can be tested.
Step 3	Evaluate the design with the interface testing techniques once the final product has been constructed.

Users do not need to consider the intricacies and complexity of using the computing system while adopting efficient HCI designs. Interfaces that are user-friendly ensure that user interactions are clear, accurate, and natural.

There are four HCI design approaches [11] that can create user-friendly, efficient, and intuitive user-interface designs:

1. *User-centered design*: This is a set of techniques that focuses on the requirements and goals of the user. When a product team creates a product, it considers the user's requirements, objectives, and feedback. Satisfying the requirements and desires of the users becomes a priority, and every design decision is evaluated in terms of whether it provides value to the users. Product designers conduct research and transform the demands of users into goals.
2. *Activity-centered design*: This is a design methodology that prioritizes the technology being utilized and focuses on how a system generates an outcome as a result of action. While the user is performing actions, the designer must map out the various tasks that must be performed on the piece of technology and discover ways to make these tasks as simple as possible to accomplish.
3. *System design*: This focuses on the components and interdependence of a system. Users adapt and use the system, while product designers represent and design controls within the system.
4. *Genius design*: The emphasis of system design is on the designer's experience, intuition, and expertise. The system's performance is validated by the user. Product designers consider and improve the system's operation.

2.10.2 Interface Testing Techniques

HCI techniques are used for a variety of objectives, including evaluating the user interface. Some techniques are given below.

- *Usability testing*: Usability testing is a technique in user-centered interaction design to examine a product by putting it to the test on real people. This is an essential usability technique since it provides direct feedback on how real users will interact with the system. It is more concerned with the design and user friendliness of the product and is tested with consumers who have no previous exposure to it. Such testing is critical to an end product's success, as a completely functional program that confuses its users will not last long. This differs from usability inspection methods, which employ a variety of techniques to assess a user interface without involving consumers.
- *Cognitive walkthrough*: The cognitive walkthrough approach is a usability evaluation technique that focuses on how easy it is for new users to complete tasks using an interactive system. The strategy is built on the concept that users prefer to learn a system by just doing tasks with it instead of studying a manual. The method is recognized for its ability to produce findings fast and

at a minimal cost, especially when compared to usability testing, as well as its potential to be used early in the design process.

■ *Heuristic evaluation*: A heuristic evaluation is a software usability inspection technique that supports the identification of usability issues in the user interface design. It requires evaluators to check the interface and grade its responsiveness to established usability guidelines. These evaluation techniques are increasingly generally accepted and performed in the media industry, where user interfaces are frequently built quickly and on a budget [12].

2.11 HCI Devices

Computers are now present in almost every aspect of our lives. In the earlier stage, batch processing techniques became popular in the nineteenth century. Human operators used punched cards to feed data into the system. These cards had holes in them and light was flashed through the holes; wherever the light passed through represented a one, otherwise it was a zero. The user-friendly interface appeared over time. The Xerox Star was the first workstation to feature a commercial GUI OS with a desktop and bit-mapped displays.

At present, the smartphone revolution has arrived. Massive computational power is contained within a portable handset. Virtual assistants such as Siri, Cortana, Google Assistant, Alexa, and Bixby use voice recognition technology to make calls, take notes, and have extended, nearly human, conversations. An analog to digital converter (ADC) converts the spoken word into a series of ones and zeros. This is then broken into fractions of a second. They are paired with phonemes, which are the smallest elements of a language. Complex statistical methods are now being investigated to determine the context of spoken words. The Internet-of-Things is the most exciting means of communication. Smartphone, laptops, cars, and homes will all be connected to a vast network of websites. Sensors at various locations will collect vital data. When the garbage cans are full, sensors will tell automated trucks to dispose of the segregated garbage [13]. Without leaving the house, one can ask a virtual assistant to turn on a car's heater.

2.12 HCI Tools and Technologies

Humans have long developed tools and technology to help them in their daily responsibilities. HCI is primarily concerned with the research of interfaces that allow people to communicate with computers, and is a field that combines computer science with behavioral science. HCI has developed from desktop computers to mobile screens. Current research focuses on integrating a voice user interface with speech recognition. Over time, this evolution has supported humans in their daily activities by minimizing effort and enhancing performance. Some of the most recent technologies, as well as their associated tools, are briefly discussed here.

2.13 HCI's Eye-Tracking Technology

Eye-tracking is a sensor technology that can detect and monitor the movements of a person's gaze in real time. Eye movements are converted into a data stream that includes eye position, gaze vector for each eye and gaze point. The technology analyses eye movements and converts them into insights. This may be used in a variety of applications. The current eye-tracking systems can be divided into four groups:

1. Eye-tracking with head stabilization
2. Remote eye-tracking
3. Mobile eye-tracking
4. Embedded or integrated systems.

2.13.1 Eye-Tracking with Head Stabilization

An eye-tracking system is made up of one or more cameras, light sources, and processing power. With the help of machine learning, powerful image-processing techniques convert the camera input into data points. Eye-tracking devices use a bite-bar or chinrest to limit the participant's head movements (Figures 2.2 and 2.3). By stabilizing the head, these devices can remove at least some head-movement abnormalities and noise from the eye-tracking data. The visual experience of participants

Figure 2.2 Employee training by eye-tracking with head stabilization.

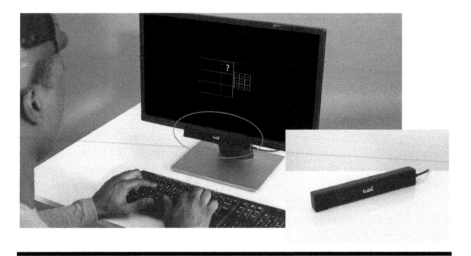

Figure 2.3 Remote eye-tracking.

is controlled by head-stabilized systems. Eye-tracking systems can be used in functional Magnetic Resonance Imaging (fMRI), magneto encephalography (MEG) and other study disciplines. With the use of eye-tracking, business administrators may help employees focus on their everyday activities. Eye-tracking technology can identify distracting industrial activities or environments. Employee training can also benefit from eye-tracking technology. It can capture an employee's visual attention while performing a task. Using eye-tracking technology and human interactions with its tools, new employees can learn everything from the perception of an expert without missing a single detail.

Limitations: The participant's comfort and natural involvement are the key limitations of head-stabilized tracking. Many head stabilization investigations do not require the participant to feel or act naturally. These systems are exclusively used in laboratory settings.

2.13.2 Remote Eye-Tracking

Modern remote systems do not require any interaction with the participant, they are termed "remote." The camera is structured to provide a distant image of the eyes and the systems can automatically adjust the camera's view point. They track eye location and head orientation using the eye center and cornea reflection remotely. While the eye-tracking device is working, a person can use a computer normally. This technology is suitable for usability testing, behavioral psychology, vision experimentation, and screen-based market research.

Limitations: Remote devices are only capable of tracking a fixed working area. Touch-screens can be challenging since the participant must frequently reach across

Figure 2.4   Mobile eye-tracking.

the camera, generating data gaps. Non-infrared light is blocked by an optical filter in most remote systems. They can work in any artificial light.

2.13.3 Mobile Eye-Tracking

Eye-tracking glasses, used in mobile eye-tracking, are also known as "head-mounted" (Figure 2.4). This type of technology usually necessitates the placement of a camera in the visual path of one (monocular) or both (binocular) eyes, as well as another camera that records the scene or field of view. This technology is used in various studies such as sports, driving, navigational, social communication, hand-eye coordination, mobile device and retail inventory testing.

Limitations: These devices, like all eye-tracking equipment, can struggle to capture eye movements in bright light. Eye-tracking cameras require an uninterrupted view of the eyes. Sometimes edges are difficult to track and mostly result in lower accuracy. When adopting mobile eye-tracking equipment, no absolute coordinate system exists.

2.13.4 Embedded or Integrated Systems

Eye-tracking devices can be integrated with other forms of devices, so aiming devices in eye surgery systems and other medical devices fall under this category. Canon has produced a number of cameras that use a gaze-based autofocus mechanism. Integrated systems are now included in virtual reality (VR) and augmented reality

Figure 2.5 Embedded system with AR and VR technology.

(AR) gadgets. Systems built into VR and AR devices can also be a control system, allowing the user to interact with information by moving their gaze [13] (Figure 2.5). Where there is no mouse or keyboard, the technology can provide an effective control method for interfaces in AR and VR.

2.14 HCI's Speech Recognition Technology

Computers can use speech recognition technology to take spoken words, understand them, and generate text from them. Speech recognition technology is already being used to boost productivity with chatbots and virtual assistants. Speech is basically a series of sound waves produced by our vocal chords. A microphone records the sound waves, which are then transformed into an electrical signal. The signal is then separated into syllables and words through modern signal processing technology. Because of recent extraordinary advances in artificial intelligence (AI) and machine learning (ML), the computer can learn and understand speech from experience over time. However, signal processing made this possible [14]. The main speech recognition tools are:

■ *Apple's Siri*: After its launch in 2011, Apple's Siri quickly became the most popular voice assistant. Siri is available on all iPhones, iPads, Apple Watches, HomePods, Mac desktops, and Apple TVs. Siri follows the user wherever they go, i.e., on the road, at home, and physically on their body. This provide Apple with a significant advantage in terms of early adoption. Siri is capable of sending a text message or making a phone call on your behalf (Figure 2.6).

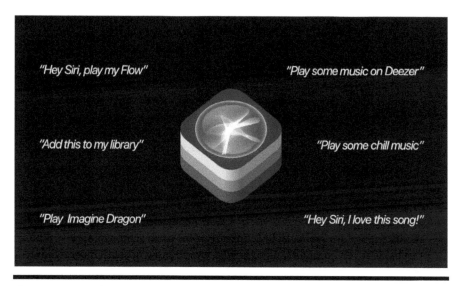

"Hey Siri, play my Flow"   "Play some music on Deezer"

"Add this to my library"   "Play some chill music"

"Play Imagine Dragon"   "Hey Siri, I love this song!"

Figure 2.6   Apple's voice assistant Siri.

Figure 2.7   Amazon's Alexa.

- *Alexa from Amazon*: Alexa and the Echo were introduced worldwide by Amazon in 2014, starting the smart speaker revolution (Figure 2.7). Alexa can be found in the Echo, Echo Show (a voice-controlled tablet), Echo Spot (a voice-controlled alarm clock), and Echo Buds headphones (Amazon's variation of Apple's AirPods). Alexa was ahead of the curve in terms of smart home device integration, including cameras, door locks, sound systems, lighting, and thermostats. Users can ask Alexa to reorder their garbage bags, and it will simply order them from Amazon. In fact, Alexa can order millions of things from Amazon without ever having to lift a finger, giving it an advantage over its competitors.
- *Google Assistant*: In early 2016, Google Home launched Alexa's most significant competitor. Google Assistant not only responds appropriately, but also

Figure 2.8   Google Assistant.

provides further context and references a source website. This is down to Google's strong search technology. Users can activate and alter vocal short-cut commands in Google Assistant to perform operations on their smartphone (Figure 2.8). This speech recognition feature is available in English and other languages as well.

2.15 The Internet of Things (IoT) Technology

IoT applications are entering into people's daily lives and practically every industry. IoT devices collect data that can help businesses gain useful insights and make important profit initiatives. From medical instruments to smartphones, smart watches to surveillance cameras, automobiles to factory manufacturing lines, all can be made smart with IoT technology. IoT technologies also contain security features that protect networked devices and their applications from internet-based cyber attacks. The IoT market is rapidly expanding, creating new business opportunities all the time. The variety of tools for developing IoT solutions is also expanding [15]. The most effective IoT software development tools are:

- *Tessel 2*: The Tessel 2 is a development board with built-in WiFi that allow users to write Node.js programs. Tessel is a networked hardware prototyping system that can be used in a variety of applications. Tessel is a JavaScript-capable microcontroller that allows users to quickly create physical devices that connect to the internet.
- *Eclipse IoT*: Eclipse IoT is an open-source community that provides a framework for developing IoT projects. This platform allows developers to propose

IoT-based solutions. Eclipse IoT is an open-source that intends to integrate popular IoT protocols, offer services and frameworks for building IoT applications and give essential tools to IoT developers. The new IoT technology can only be effectively applied if it is used in an open-source approach. Eclipse provides resource management solutions for developers, frameworks, and services.

■ *Arduino*: Arduino is a smart choice if you want to develop a computer that can sense and control the physical world better than a traditional stand-alone computer. Arduino is an open-source prototype and easy-to-use IoT platform that offers a perfect blend of IoT hardware and software, through an interactive electronics platform that uses a set of hardware specifications. The software includes the Arduino programming language and the Arduino Integrated Development Environment (IDE).

■ *PlatformIO*: PlatformIO Core and PlatformIO IDE are both available for personal and commercial use for free. The user pays for access to additional features. This platform includes a build system, as well as library management and an IDE. Users can either port the IDE to the Atom editor or install it as a plug-in. PlatformIO's best feature is that it works with over 200 different boards. PlatformIO is a serial port monitor that comes with fantastic debugging integration.

■ *IBM Watson*: IBM Watson is an application programming interface (API) that allows users to integrate a variety of cognitive computing technologies into their IoT applications. This is a unique technology that can be used to forecast the future. IBM Watson's multiple capabilities assist in the development of chatbots that can interpret natural language, simplifying the responsibilities of IoT developers. These chatbots can then be launched on messaging systems and websites that are accessible via a variety of devices. IoT developers use IBM Watson to construct cognitive search and content analytics engines quickly and successfully.

■ *Raspbian*: IoT tech enthusiasts developed this IoT IDE on the Raspberry Pi board. It is a vital IoT development tool with over 35,000 packages with several examples, as well as speedy installation via pre-compiled software. Another great feature of this tool is that it is constantly being improved and expanding the computational reach so that users can get the most out of it.

■ *OpenSCADA*: The Eclipse IoT industry groups developed this tool as part of their SCADA project. It is platform-independent and is famous for its security, flexibility, and modern design. OpenSCADA also includes front-end apps, back-end apps, libraries, set-up tools and interface implementations, as well as support for editing and debugging. Its various statistical tools can be used in the creation of complex IoT applications.

■ *Kinoma Create*: This is a device that can connect two devices without requiring extensive JavaScript skills. Kinoma Create includes everything needed to build simple IoT applications, such as connecting lights, temperature, or movement sensors for a specified purpose and receiving mobile alerts in the case of any changes. Kinoma's website also has a number of tutorials on its practical applications.

■ *Device Hive*: Device Hive is a free and open-source machine-to-machine (M2M) communication framework that was released in 2012. It is built on Data Art's AllJoyn and is one of the most popular IoT app development platforms. It's usually a cloud-based API that you can control remotely without having to set up a network. Libraries, portals, and management protocols are all included. Security, automation, smart home technology, and sensors are the most common uses. It also features a growing community and a variety of online services to support its users.

2.16 HCI's Cloud Computing Technology

Cloud computing is a virtualization-based technology that enables us to create, configure, and personalize applications over the internet. A development platform, hard disc, software program, and database are all included in the cloud technology. A cloud is a network or the internet. Rather than using local storage, it uses remote servers on the internet to store, manage, and access data. Data can include files, photos, documents, music, video, and other types of media. Small and large businesses worldwide use cloud computing technology to store data on the cloud and access it from anywhere, at any time, via an internet connection. Service-oriented architecture and event-driven architecture are integrated into the cloud computing architecture [16, 17]. The following is a list of cloud computing technologies:

1. *Virtualization*: Virtualization is the means of producing a virtual environment on a single server to run various applications and operating systems. The virtual environment can be any combination of operating systems, storage devices, network application servers, and other environments. In cloud computing, the concept of virtualization enhances the implementation of virtual machines. A virtual machine is a software program that may act as a real computer and execute functions such as running apps or processes on demand.

2. *Service-oriented architecture (SOA)*: SOA enables enterprises to use on-demand cloud-based computing solutions to adapt to varying business needs. It can be used with or without cloud computing. The benefits of SOA include ease of maintenance, platform independence, and scalability. The two primary roles in SOA are Service Provider and Service Consumer.

3. *Grid computing*: Distributed computing is another term for grid computing. This is a processor architecture that integrates numerous computing resources

from many locations in order to achieve a common goal. In grid computing, parallel nodes join the grid to form a computer cluster. These computer clusters vary in size and are capable of running on any operating system. Grid computing is most commonly used in ATMs, back-end infrastructures, and market research.

4. *Utility computing*: Utility computing is the greatest IT service model right now. It uses a pay-per-use model to deliver on-demand computing resources, such as computation, storage, and programming services through an API. It reduces associated costs while maximizing resource efficiency. Utility computing has the benefit of increasing flexibility and making management easier. Large corporations like Google and Amazon have developed their own utility services for compute storage and applications.

These four technologies are widely used in a variety of applications. The following are cloud computing tools:

- *Amazon Cloudwatch*: Amazon Web Services provides cloud resource and application monitoring. It provides a summary of the system's entire activity and well-being. This system has a lot of potential for optimizing processes. The best thing about cloud services is that users don't have to install any software or pay for expensive installation services.
- *Cloudability*: This is a financial management solution for tracking and monitoring all cloud expenses across an enterprise. It assists in identifying cost-cutting opportunities, preparing reports, and providing budget alert and recommendations via SMS and gmail.
- *Cloudyn*: This tool was created to keep corporate IT from overspending for Amazon cloud services. Cloudyn's services provide users with a dashboard that displays complete information about all of their virtual machine models, databases, and storage. Cloudyn also offers insight into failed suggestions and solutions. Cloudyn features include intuitive dashboards, overall cost analysis, usage breakdown analysis, resource cost analysis, usage trends analysis, financial projections analysis, and unused resource elimination recommendations.
- *Informatica*: Informatica is a widely used data processing tool to extract, transform, and load (ETL) data. It gathers information from many sources and delivers reliable results and automates deployment. Multiple operations are executed simultaneously. It has a centralized cloud server that makes data easy to access and track. Maintaining and monitoring data are simple.
- *Chef*: Opscode offers an open-source Ruby-based configuration management tool under the Apache license. Administrators can programmatically configure virtual systems and eliminate all repetitious manual tasks with the help of a hosted Chef cloud system.

- *Enstratius*: Enstratius is a cloud computing solution that provides cross-stage cloud infrastructure for private, public, and hybrid clouds, depending on the needs of an organization. It enables self-service cloud resource provisioning and deprovisioning. Users can manage all cloud resources with a single login and support enterprise authentication solutions like OpenID and SAML 2.0.
- *Microsoft Azure Automation*: Azure Automation is a cloud-based automation and configuration service that allows users to manage their Azure and non-Azure environments consistently. It includes process automation, update management, and configuration capabilities that will help users reduce errors and minimize the time it takes to build infrastructure. Azure Automation also allows for automated maintenance and compliance monitoring. Azure could be a good option if a user needs to run a variety of operating systems.

Cloud computing is a game-changing technology that has enabled businesses to supply products and services while also dealing with cyber-security challenges, massive data management, and quality control. Cloud computing solutions have been used by organizations, ranging from startups to multinational corporations to create applications and automate business processes.

2.17 HCI Applications

HCI is important since it will be required for products to be more effective, secure, useful, and efficient [18]. It will improve the user's experience in the long term. Some of the most useful applications are.

- *In daily life*: Technology has now penetrated every aspect of human life. Even if a person does not own or use a computer directly, computers still have an impact on their lives. ATM machines, train ticket vending machines, and hot beverage vending machines are just a few examples of computer interfaces that individuals can use on a daily basis without owning a computer. HCI concepts should be investigated and properly considered when developing any of these systems or interfaces, whether it is for an ATM or a desktop PC, to ensure that the interface is comfortable, helpful, and effective.
- *The commercial sector*: HCI is essential for every business that uses computers in its daily operations. Computers are used by employees in their daily activities. They may use distinct software for their operations. If the design is poor, they will experience frustration and other issues as well. HCI is especially important when designing safety-critical systems, such as power plants and air traffic control centers. Design errors can have severe implications in certain situations, including the death of the user.
- *Assistive technology*: HCI is an important concern when selecting systems that are not only effective but also accessible to people with disabilities. The core

concept of HCI is to provide safe, effective, and efficient systems to everyone, including people with disabilities. Each system developed using HCI user-centered methodologies and principles will be as accessible to disabled individuals as possible.

■ *Healthcare*: Computers have become a significant component of hospitals, labs, and pharmacies. In hospitals, they're used to keep track of patients and medicine. They are also used to scan and diagnose certain disorders. Computerized equipment also performs ECGs, EEGs, ultrasounds, and CT scans, among many other procedures.

■ *Government*: In government services, HCI plays a vital role. Budgets, sales tax departments, income tax departments, male/female ratio computations, and so on are some of the important uses in this category. Voter list computerization, PAN card computerization, and enhanced weather forecasting are other applications of HCI.

2.18 Advantages of HCI

HCI is a field that cannot be avoided because it has now been integrated into every phase of product development. There are numerous advantages of HCI and some of them are:

■ Users learnt how to communicate with computers using the keyboard, mouse, and desktop when computers first arrived. Users started with Microsoft Excel for calculations, Word for documents, PowerPoint presentations, and virtual desktops via GUIs, but GUIs have revolutionized the way computers and users interact.

■ HCI improves data security.

■ Mobile phones are the most widely used medium of communication in the world.

■ Music players, global positioning system (GPS), news apps, calendars, games, and other applications are all available on today's smartphones.

■ HCI is adaptable, versatile, and simple to use.

■ AR and VR are making games more fascinating.

2.19 Disadvantages of HCI

HCI can be very useful, but there are also certain disadvantages:

■ It is necessary to instruct the user in what to do.

■ Depending on the nature of the activity, even a tiny human error might have a significant influence on cost, safety, and other critical considerations.

■ Sometimes people are unable to communicate with computer interfaces due to problems, such as memory loss, a lack of management capability, and cognitive impairments.
■ Attacks by viruses, and deliberate hacker attacks.

2.20 Conclusion

The study and deliberate design of human and computer interactions are known as human-computer interaction (HCI). HCI is used in a variety of computer systems, such as in offices, nuclear processing, air traffic control, and video games. HCI systems are simple, secure, efficient, and interesting. There is a vast body of knowledge on HCI, comprised of concepts, rules, and design standards. In this present technological age, cleverly designed computer interfaces encourage the use of digital devices. HCI allows two-way communication between man and machine. Because of the effectiveness of the communication, people assume they are interacting with human personalities rather than a complicated computational system.

Nonetheless, creating an appropriate interface at the first try is impossible. This will be a huge challenge in the coming years. Tasks, users, needs, context analysis, and interface design must all have more explicit, direct, and formal relationships. Current ergonomic information should also be used more directly and simply in interface design. As a result, it is essential to provide a solid HCI foundation that will effect future applications, such as targeted marketing, eldercare, and even psychological trauma healing.

References

1. Wilbert O. Galitz (2007) "The Essential Guide to User Interface Design." DreamaTec. Available at: https://mrcet.com/pdf/Lab%20Manuals/IT/ R15A0562%20HCI.pdf
2. Khalid Majrashi, and Margaret Hamilton (2014) "User Experience of University Websites." Available at: www.researchgate.net/publication/295857951_User_Experi ence_of_University_Websites
3. Jenny Preece (1994) "Components of HCI." Available at: http://ir.lib.cyut.edu.tw:8080/ bitstream/310901800/14308/3/c2.pdf
4. Constantine Stephanidis, Gavriel Salvendy, Margherita Antona, Jessie Y. C. Chen, Jianming Dong, and Vincent G. Duffy (2019) "Seven HCI Grand Challenges." https:// doi.org/10.1080/10447318.2019.1619259
5. Brad A. Myers (1993) "Why Are Human-Computer Interfaces Difficult to Design and Implement?" Available at: www.cs.cmu.edu/~bam/papers/WhyHard-CMU-CS-93-183.pdf
6. Lowella Viray (2016) "Exploring Human-Computer Interaction: Its Issues and Challenges." Available at: www.academia.edu/38324061/Exploring_human_com puter_interaction_its_issues_and_challenges

7. Fakhreddine Karray et al. (2008) "Cooperative Multi Target Tracking Using Multi Sensor Network." *International Journal on Smart Sensing and Intelligent Systems*, 1(3).

8. Educative (2021) "Introduction to HCI and Design Principles." Blog. Available at: www.educative.io, www.educative.io/blog/intro-human-computer-interaction (accessed May 31, 2022).

9. The Interaction Design Foundation (2014) "The 5 Main Characteristics of Interaction Design." Available at: www.interaction-design.org/literature/article/the-5-main-characteristics-of-interaction-design (accessed August 1, 2015).

10. Educative (n.d.) "Introduction to Human-Computer Interaction & Design Principles: Interactive Courses for Software Developers." Available at: www.educative.io, www.educative.io/blog/intro-human-computer-interaction (accessed May 31, 2022).

11. Alma Leona (2016) "Design Methods and Methodologies in HCI/ID." Available at: www.uio.no/studier/emner/matnat/ifi/INF2260/h16/lec2_designmethods.pdf

12. Wikipedia (n.d.) "Human–Computer Interaction." Available at: https://en.wikipedia.org/wiki/Human%E2%80%93computer_interaction (accessed August 1. 2021).

13. Bitbrain (n.d.) "Different Kinds of Eye Tracking Devices." Blog. Available at: www.bitbrain.com/blog/eye-tracking-device (accessed June 12, 2020).

14. Summa Linguae (2021) "The Complete Guide to Speech Recognition Technology." Available at: https://summalinguae.com/language-technology/guide-to-speech-recognition-technology/ (accessed June 12, 2021).

15. Eduonix (2019) "Top 10 Popular IoT Development Tools." Blog, May 14. Available at: https://blog.eduonix.com/internet-of-things/top-10-popular-iot-development-tools/

16. Javatpoint (n.d.) "Cloud Computing Technologies." Available at: www.javatpoint.com/cloud-computing-technologies (accessed May 31, 2022).

17. Jigsaw Academy (n.d.) Blog. Available at: www.Jigsawacademy.Com/Blogs/Cloud-Computing/Cloud-Computing-Tools/ (accessed May 31, 2022).

18. Utsav Mishra, and AnalyticsSteps (2021) "Human-Computer Interaction (HCI): Importance and Applications." Blog. Available at: www.analyticssteps.com/blogs/human-computer-interactionhci-importance-and-applications

Chapter 3

Edge/Fog Computing: An Overview and Insight into Research Directions

Priya Thomas and Deepa V. Jose

3.1 Introduction

Cloud computing has emerged as a platform which provides on demand access to computing resources and services anywhere, at any time. It enables remote access for storage, software, data repositories, etc. and offers services such as networking, data filtering, and analytics. The cloud services can be accessed using the pay per use policy, making it more flexible and customer-friendly. There are lots of cloud providers, such as AWS, Google Cloud, Microsoft Azure, etc., that guarantee seamless services to end users. The reliability of the cloud technology will greatly reduce the infrastructure and maintenance costs, making it very popular.

The cloud computing technology can be classified into public, private, and hybrid cloud, based on the access model followed. The *public cloud* offers services to users by resource sharing. The users can access the services of the remote cloud without building their own infrastructure. Data availability depends on the underlying network condition, and the security and privacy policies will be defined by the cloud providers. The *private cloud* provides dedicated infrastructure to users with more security as they are built within the premises of an organization. It helps to access data faster and with greater reliability. *Hybrid clouds* are a combination of private and public clouds which can be relied on to improve the scalability. This type helps to access all the benefits of the public cloud without compromising security and speed.

DOI: 10.1201/9781003424550-3

The cloud computing services can be classified into the following:

- *Infrastructure as a Service (IaaS)*, which provides services such as remote access to storage, computing, virtual machines and networking resources.
- *Software as a Service (SaaS)*, which helps to access software remotely without directly installing them on PCs.
- *Platform as a Service (PaaS)*, which helps to access computing platforms, such as operating systems, program execution environments, databases, etc. remotely.

All these services are available to cloud users, with choice of service based on their requirements.

Cloud computing offers numerous benefits to its users, such as scalability, agility, elasticity, low cost, high performance, higher productivity, etc. Cloud computing also suffers from a few drawbacks [1], such as single point of failure, security issues, high response time, longer processing delays, latency, less reliability, etc. Research is continuously going on to improve the features of cloud computing models. Among the different solutions, one of the most trustworthy solutions is the incorporation of fog/edge computing. The terms fog and edge are usually used interchangeably in the literature, and these two types of computing help to improve the performance of applications which depend on cloud computing technology.

The term fog computing was originally coined by Cisco. The concept of the edge/fog was introduced to bring programming to the edge of the network instead of remote data centres. The major drawback faced by cloud computing was related to high response time due to the longer distance travelled by the data. The cloud data centres reside in geographically distant locations from the user or application which uses it. This significantly increases the response time and the delay will be much more if the network is congested, making cloud a less promising solution for delay-sensitive applications. Edge/fog computing helps to reduce this problem by reducing the distance between the end user and the computing nodes. Even though the terms are used synonymously, there is a difference between edge and fog. The difference mainly lies in the location where the data is processed.

In edge computing, the data is processed on the edge of the network. For example, if the Internet of Things (IoT) is an application that implements edge computing platform, the IoT sensor itself can be configured as an edge node and data can be processed in this node. In fog computing, the processing happens in processors connected to a local area network (LAN) or the LAN hardware itself which is nearer to the end user. So, both of these computing platforms help to process data locally and respond faster to the application. The time-sensitive data will be processed in these edge/fog nodes and the response will be transferred immediately to the end devices, thereby reducing latency and delay considerably. The less time-sensitive data will be transferred to the remote cloud for further processing, forming a three-layered architecture.

The introduction of edge/fog computing has greatly reduced the delay in getting a response in time-sensitive applications [2]. The edge/fog layer is not an independent entity. It relies on the cloud for further processing and permanent storage of data. Thus, the end devices, the edge/fog layer and the cloud together form the design architecture for many applications. Like any technology, the edge/fog also has a few drawbacks which are an area where active research is still going on. The recent research shows that the latency, response time, delay, etc. can be considerably reduced by incorporating edge/fog computing, whereas the impact on other parameters needs to be tested. The lack of a service-level agreement (SLA), offloading issues, resource monitoring, etc. need to be addressed to make the system much better. The major challenges to be addressed on edge/fog incorporation are discussed in later sections.

In short, cloud computing helps to reduce the infrastructure and maintenance costs by providing on-demand access to a pool of resources, including software, databases, storage, etc. and provides numerous services to users based on a pay per use policy. One of the major drawbacks of cloud computing lies in the delay in getting a response as the data centres are in geographically remote locations. The cloud also suffers from security and privacy issues and higher bandwidth consumption. The introduction of edge/fog computing solves many issues related to cloud implementation. Section 3.2 gives an introduction to edge/fog computing, defines the architecture, the benefits of edge/cloud integration, applications and discusses research challenges to be addressed when integrating edge/fog to the cloud environment.

3.2 Edge/Fog Computing Basics

The explosive growth of technology has resulted in newer innovations, such as the IoT, which continuously generate millions of data. The IoT greatly relies on the cloud for storage and processing due to its power-constrained nature. The large volume of data which is transferred back and forth in cloud data centres creates network congestion and increased network bandwidth utilization. This also creates prolonged delay in getting a response which seriously affects the performance of IoT devices. Edge/fog computing brings processing nearer to the IoT devices, thereby reducing latency and delay.

3.2.1 Edge Computing

Edge computing refers to performing the processing at the edge of the network, which can be either in the IoT sensor itself or the gateways [3]. The time-sensitive data will be analysed and processed in the network's edge and the response will be immediately sent to the end devices. Many companies are making the shift from cloud to edge or have incorporated the edge as a middle layer to improve the performance of the existing applications.

The edge nodes are designed to be capable of performing computational tasks locally. Apart from this, edge devices are configured to request and respond to data from remote cloud data centres. Edge nodes can perform data storage and analysis, computational offloading, processing, and request and response services, as well as distribute request and delivery services from remote cloud data centres to the user. As edge nodes are capable of doing these tasks, they should be designed in such a way that reliability, usability, privacy and security should be top priority. The architecture must also support requirements such as differentiation, extensibility, isolation, and reliability.

Edge computing has numerous definitions in the literature. Edge computing is an autonomous computing model which includes a set of heterogeneous devices arranged in a distributed manner that communicate with the underlying network and perform computing tasks, such as data storage, analytics and processing [4]. The edge computing can be implemented as a lease-based service, if required, without needing to build the necessary infrastructure. Mobile edge computing is a similar term to edge computing, which helps mobile users to access the features of edge computing. Terms such as Cloudlets, MicroCloud, fog, etc. are terms similar to edge computing which are designed as subsidiaries to cloud technology.

3.2.2 Fog Computing

Fog computing is synonymous with edge computing with the difference being only in location where the data is processed. The fog follows a distributed architecture where the incoming tasks are offloaded to different fog nodes [2]. In fog computing, the processing happens in LAN hardware or processors attached to a LAN. This also greatly reduces the response time and reduces latency issues.

According to Cisco, fog computing can be defined as an extension of the cloud computing paradigm from the core of the network to the edge of the network. It is a highly virtualized platform which supports processing, analytics, data storage, and networking services between end devices and cloud servers. The fog services are usually evaluated using the Quality of Service (QoS) metric. The QoS is usually evaluated based on four parameters: (1) connectivity; (2) reliability; (3) capacity; and (4) delay.

Connectivity refers to the ability to stay connected to the network using the appropriate protocols. As the fog follows a distributed nature, failure of one fog node will not affect the connectivity. Reliability is the measure of how error-free the data is. Fog computing performs processing in close proximity to the end devices. This will considerably reduce the errors and issues which can happen for data which needs to travel long distances. This ensures reliability in fog computing.

Capacity is defined by bandwidth usage and storage. Capacity needs can be improved based on how well the data is offloaded to different fog nodes and how well the filtering is performed and the cloud resources are utilized. Delay is the time between the request and the response, which also affects the QoS requirement of fog computing. As processing happens on the edge of the network, latency will be reduced and the response time will be shorter, which considerably reduces delay and improves performance.

3.2.3 Understanding the Differences between Edge and Fog Computing

Both edge and fog computing are designed as the subsidiaries to cloud computing. Even though both terms are used interchangeably, there exist a few differences between them.

The main difference between the edge and fog environments is [4], in edge computing, computation happens usually on end devices. It can be any IoT sensor or gateways which are close to the sensors. Fog computing shifts computing to the LAN hardware or processors connected to the LAN, which is near to the end devices but at a longer distance away compared to the edge devices. Edge computing is closer to the end devices compared to fog computing. This helps to further reduce the transmission time. Thus, edge computing generates a response with less delay and latency.

Edge computing has less capability to perform data filtering and analysis. It processes all the data and generates a response without analysing the requirement. Fog computing offers an environment to test and analyse the incoming requests before processing. Memory and storage space can be saved by enabling fog computing. The fog computing supports data filtering where only the required data will be transferred to the cloud, which helps to reduce bandwidth consumption. The cloud will not be overloaded with millions of bits of data. The network congestion, jitter, delay, etc. can be reduced with fog implementation.

The edge computing environment is less scalable than fog computing. The edge nodes are usually part of an IoT network. Fog nodes are configured in the LAN hardware and have a strong infrastructure support, making the fog easily expandable. The fog environment is capable of incorporating newer nodes, making the technology more scalable.

Edge computing is considered a part of the IoT network itself. The data is processed within the IoT network itself, making the data more secure. The fog nodes capture the data from nearby end devices and process it. The data has to travel at longer distance, which increases the chances of attacks, compared to edge computing. But when considering the cloud processing, fog computing offers better security and less chance of errors. Fog nodes are considered a part of the cloud computing infrastructure itself.

The edge/fog technology has only minor differences among them. But when compared to the cloud technology, we can see that differences are major. Table 3.1 depicts the differences between the cloud, fog and edge computing models. Table 3.1

Table 3.1 Comparison Between Cloud Computing, Edge Computing, and Fog Computing

Requirement	Cloud computing	Edge computing	Fog computing
Latency	High	Low	Low
Delay jitter	High	Very low	Very low
Location of server nodes	Within the Internet	Edge of network	At the edge of the local network
Distance between the client and server	Multiple hops	One or two hop	One hop
Security	Undefined	Can be defined	Can be defined
Attack on data en route	High probability	Low probability	Very low probability
Location awareness	No	Yes	Yes
Geographical distribution	Centralized	Distributed	Distributed

clearly shows that response time, delay and jitter will be comparatively less for edge/fog. The security is usually predictable and can be defined in the edge/fog whereas it is undefined in the cloud. The chance of an attack on the data en route is much less in fog computing and less for edge computing. But cloud computing has more chance of attacks. The cloud also does not support location awareness and mobility. The cloud is designed as a centralized data centre whereas both edge and fog follow a distributed pattern.

3.2.4 Edge/Fog Computing Characteristics

Edge/fog computing has many characteristics [5], which are similar to the cloud computing. However, there are some characteristics which differentiate cloud computing from edge/fog computing,

3.2.4.1 Low Latency

One of the most promising characteristics of edge computing is latency reduction. The latency can be defined as a delay in communication in simple terms. It refers to the overall delay happening for packets from the source to the destination, right from packet transfer by the sender to decoding at the receiver side. As the edge computing performs computations at the edge of the network, latency can be considerably reduced compared to cloud implementation.

3.2.4.2 Less Bandwidth Consumption

The large amount of data generated by different applications consumes lots of bandwidth while travelling from source to remote cloud data centre. This can be reduced by incorporating edge devices, as most of the time-constrained requests can be processed locally by the edge computing. The edge/fog can also filter data and perform data analytics and forward only the required data to the remote cloud. This greatly reduces the bandwidth consumption, making it preferable.

3.2.4.3 Mobility Support

Fog/edge computing supports mobility with the help of protocols, such as Locator ID Separation Protocol (LISP). LISP is an RFC 6830 standard protocol developed by the Internet Engineering Task Force LISP Working Group. The LISP protocol decodes the location information from the host packet and provides the required mobility support for applications using edge computing.

3.2.4.4 Heterogeneity

Heterogeneity refers to different types of elements and technologies involved in the edge computing infrastructure. The edge computing elements include the end devices, servers and the edge network which performs a variety of functions. The APIs, communication protocols, policies and platforms involved also contribute to the heterogeneous nature of edge computing.

3.2.4.5 Location Awareness

Applications which require location awareness such as transportation and utility management require edge/fog computing environment. Cloud computing does not support geographically based applications due to lack of the location awareness feature. The edge/fog helps to collect and process information based on geographic location.

3.2.5 Benefits of Edge/Fog Computing

3.2.5.1 Cost

Edge/fog servers reduce the cost of building a cloud-like infrastructure. The maintenance of edge servers is also easier, compared to the cloud. Companies can easily set up an edge computing environment and can monitor and maintain it with comparatively lower costs. The implementation as well as the maintenance of the edge/fog can bring lower costs.

3.2.5.2 Speed

The edge/fog computing services are provided at the edge of the network which increases the speed in communication. The incoming requests are processed faster and the response is generated immediately. The data which needs detailed processing only will be transferred to the remote cloud. This will reduce the traffic in the network connecting edge and the cloud also. This integration greatly enhances the speed and performance of the system.

3.2.5.3 Scalability

The edge computing infrastructure can be scaled as per the requirements. Companies can easily expand the edge infrastructure in their premises as needed. The fog computing which is more scalable than edge can be used for organizations which needs to scale their resources.

3.2.5.4 Performance

The performance of applications which rely on edge/fog computing improves as the delay, jitter, response time, etc. will be reduced. The faster processing and filtering done at the edge level contribute to high performance. Proper task scheduling algorithms should be chosen to distribute the load among the available fog nodes. This also will enhance the performance of the applications.

3.2.5.5 Security

The edge computing offers better security than the cloud as the edge/fog implementation lies within the premises and is under the control of the company. The company can set up the required policies and can decide which data has to be passed to the remote cloud and which has to be processed locally, which enhances security. The data need not travel longer distances, reducing the chance of data manipulation. The longer the transmission time, the more the chance of attacks and data loss. The edge/fog reduces this risk and thus contributes to security.

3.2.5.6 Reliability

Edge/fog computing offers a more reliable service compared to the cloud. The cloud-enabled applications completely rely on the remote cloud for storage, data access and data processing. A failure or crash of the cloud server makes the cloud services completely inaccessible. Fog and edge computing models are distributed in nature, reducing the problems arising from a single point of failure. The failure of one node will not affect the other nodes and the data transfer. The failed node can be identified and tasks can be rescheduled to compensate the loss. The data can be monitored properly to avoid errors creeping in, making the system more reliable and trustworthy.

3.2.5.7 Agility

The fog/edge computing model can be built by small and individual businesses, making it more agile. The system deployment is now possible with the help of open software interfaces and development kits. There are numerous edge/fog simulation platforms which can be explored to test the use cases.

The edge/fog devices can easily be configured and require lower infrastructure costs for operation and maintenance. The edge/fog cooperates with the cloud data centre and follows a layered design approach. The available architectural pattern for edge/fog incorporation and its features are discussed next.

3.3 Architecture of the Edge/Fog Computing

The edge/fog computing has the advantage of reducing latency and response time and increasing the performance of time-sensitive applications. But the edge/fog cannot be considered a replacement for cloud computing. They complement each other, giving rise to a three-tier architecture [6]. The different layers include end devices at the lower layer, followed by the edge/fog layer, followed by the cloud layer. All three layers coordinate with each other to offer the services. The different layers are enriched with a set of devices and tools which synchronize together to increase the performance of the system. Different architectural patterns are available in the literature. The components of each layer and services provided by each layer are areas where active research is in progress. The most common architecture is the three-layered one. It serves as the basis for the different models currently available. There are numerous simulation tools which support the three-tier architecture. The layout of the cloud-edge/fog architecture is demonstrated in Figure 3.1.

The lowermost layers include different sensors, actuators, RFID devices, mobile devices, etc., which continuously generate some form of data. The data will be transferred to the middle layer, i.e., the edge/fog layer. This layer will filter the data, process it and send a response back to the end devices. This layer will also pass the data to the cloud layer for further processing and permanent storage. The cloud layer will process the data, do proper analytics and send a response back. It also keeps a copy of the processed data. The end devices, on receiving a response either from the cloud or the edge, will proceed to the next stage. This is how all three layers coordinate together to accomplish any task.

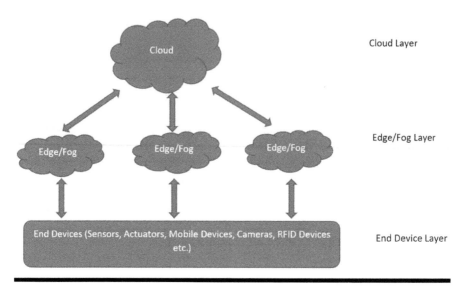

Figure 3.1   Cloud-fog/edge architecture.

3.3.1 Basic Components of the Three-Tier Architecture

3.3.1.1 End Device Layer/Physical Layer

The physical and virtual nodes are the end devices involved in the lower layer of the architecture. End devices will have requests to send to the middle fog layer, which needs to be processed, and a response should be gathered [7]. Physical and virtual nodes communicate with the fog layer through a wired or a wireless medium. Millions of bits of data are generated by the end devices at any time. Different IoT-enabled devices, smart devices, mobile phones, RFID devices, etc. are part of the lower layer. Communication policies and security standards have to be enforced to enable a smooth transfer of the data between the physical layer and the fog layer.

3.3.1.2 Fog/Edge Layer

The fog layer includes the fog nodes, fog devices and fog servers in general. Fog nodes are the devices which are configured to collect data from the virtual and physical nodes. The fog nodes include fog servers where the data collected is stored and analysed. Fog servers will be connected to fog devices which act as an intermediate device between the end devices and the servers. The fog devices need to be connected to the fog servers and this is enabled using gateways [8].

The fog servers will be responsible for performing the offloading and storing various algorithms which decide the offloading strategy. Offloading refers to the distribution of incoming requests to the available fog nodes. The fog layer also includes

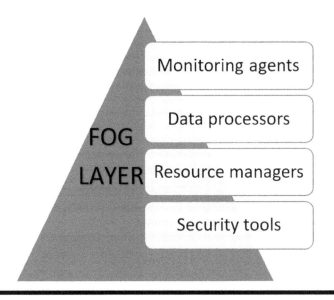

Figure 3.2   Entities of the fog layer.

sophisticated agents for data analytics, processing, filtering, device management, application life cycle management, etc. The different features of fog computing are implemented using components, such as monitoring agents, data processors, resource managers and security agents. Figure 3.2 presents the main entities of the fog layer.

Monitoring agents in the fog layer are responsible for continuously monitoring and evaluating the performance of the connected devices and programs. The monitoring services will use a set of APIs to ensure that resources are properly utilized. The communication services, resource availability, offloading policies, system work-load, etc. will be continuously monitored.

Data processors, which are part of the fog layer, are software which performs the filtering, analysis and error detection and correction. They run on fog nodes and decide whether incoming data needs to be forwarded to the remote cloud or should be processed in the fog node. The load distribution, processing of data, embedding intelligence to analysed data, etc. are some features of the data processors.

The resource manager ensures the smooth functioning of different fog devices in the fog layer and end device layer. The resources should be properly allocated and data has to be transferred to the appropriate devices with proper synchronization. The data back-up has to be done in a timely fashion and resource requirements should be properly monitored. The resource manager should also remove the chances of data redundancy.

The fog layer includes a set of security tools to check and authenticate the data source and to prevent unauthorized access. The raw data has to be processed without any errors and communications have to be properly encoded. The security policies for the connected applications and users haves to be properly framed and should be monitored strictly, using security tools to maintain the reliability of the data.

The different units in the fog/edge layer cooperate with each other and work in synchrony with the upper cloud layer and lower physical layer.

3.3.1.3 Cloud Layer

The filtered data from the middle layer will be forwarded to the upper cloud layer for permanent storage. The cloud performs detailed analytics and processing which cannot be done in the fog layer. The cloud is rich in resources and act as a supporting layer for the intermediate edge/fog layer. The cloud layer provides the permanent storage of data and data accessibility at any time.

The cloud layer includes components such as client interface, which includes a set of graphical user interfaces (GUIs) which help applications to interact with the cloud, applications which include software capable of delivering the required functionalities for requesting client models and a wide range of services models, such as Software as a Service, Platform as a Service, Infrastructure as a Service, etc. The cloud has a large volume of storage space and proper security principles to ensure the safety of the data. The management module in the cloud is responsible for managing the front end and back end and also checks resource utilization and availability. The management module ensures that communication happens without errors and also monitors the performance of the system. The cloud storage and services can be accessed remotely. The major components of the cloud layer work together to ensure data transfer between the lower layers. The response time depends on availability and network conditions. But as the edge/fog are capable of performing most of the processing locally, the performance of connected applications can be guaranteed. The different units in the cloud layer which supports data manipulation are shown in Figure 3.3.

The major components of the sub-layers of the three-tier architecture coordinate together to improve the performance of the connected system. The working of the model can be summarized as follows. The lowermost end user layer continuously generates data which moves to the edge/fog layer through the communication medium. The fog devices in the middle layer accept the data and perform analysis. The time-sensitive data will be processed in a faster mode to reduce the response time. The response generated will be sent back to the requesting devices in the lower layer. The edge/fog layer has data processors which decide whether the data has to be offloaded to fog nodes or to the remote cloud [7]. If the data is not time-critical, it will be filtered and sent to the remote cloud for further storage and processing. The cloud layer, after processing the request, will send the response back to the lower layer. Thus, the combined strategy helps to reduce the delay without compromising the performance and resource requirements.

The above model has numerous applications in different sectors. Most of the IoT applications are currently shifting from cloud-only infrastructure to edge/

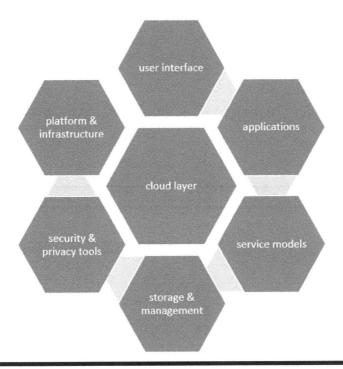

Figure 3.3   Components of the cloud layer.

fog–cloud integration to reduce latency. The research in this area is active and different algorithms are being tested to make the applications more productive. The different layers are independently analysed and problems are identified. The research initiatives in this area are gaining popularity nowadays.

3.4 Applications of Edge/Fog Computing

The edge/fog computing platform supports numerous applications. The different applications areas include healthcare, IoT-enabled applications, such as smart cities, smart grid, smart vehicles, etc., banking and the finance industry, augmented reality, etc. [4]. Some application use cases of edge computing are now discussed.

3.4.1 Healthcare Industry

The healthcare sector generates a vast amount of data every second, which needs to be stored, processed and analysed. The analysis and decision-making process of health information are highly time-critical and cannot tolerate delay. Edge or fog computing reduces the delay by providing faster response times. The data can be filtered and transferred to the cloud for permanent storage.

3.4.2 IoT Applications

IoT applications depend on the cloud for storage and processing as the IoT nodes are resource-constrained. The sensors generate millions of bits of data which need to be stored and processed at a fast rate. Cloud computing models suffer from huge delays as the bulk amount of raw data needs to be filtered and analysed before generating a response. Edge/fog computing helps data filtering and reduces delay as the processing happens near the IoT devices. The bandwidth consumption, delay and traffic congestion will be reduced after the integration of edge and fog computing.

3.4.3 Augmented Reality (AR)

Augmented reality (AR) is the technique of enhancing the existing physical entity using system-generated inputs. The technique should have minimum delay and a high data handling rate to generate the correct information [9]. Even small delays in getting the response may damage the client's skills. Therefore, edge/fog computing has a great impact on the applications supporting augmented reality.

3.4.4 Banking and Finance Industry

The banking and finance industry place great importance on data security and privacy. Sensitive information should not travel a long distance to reach the remote cloud, increasing the chance of data being compromised. Edge/fog computing models can be built within the different banking branches locally. The processing can be done locally, ensuring data privacy and security. Fraudulent actions can be analysed immediately and appropriate actions can be taken without delay by incorporating edge/fog computing in the banking and finance sector.

3.4.5 Manufacturing Industries

Manufacturing industries require faster response as their operations are highly time-critical. The delay in getting the response from the remote cloud can result in on-site accidents and errors. The machines generate bulk amounts of raw data which need to be analysed to derive patterns and decisions. The huge volume of data will overload the cloud server, reducing its capacity to generate a response. Edge/fog computing helps by data filtering, rather than passing massive amount of raw data to the central server. This makes the decision-making process easier. The edge also helps in processing data locally and reducing the response time and transportation cost for time-critical applications.

3.4.6 Automobile Industry

EdgeAI is a promising platform for the automobile industry. The automobile industry is rapidly growing. Edge computing provides the required support for predictive maintenance, assists drivers in identifying problems and taking emergency control, driver monitoring to prevent accidents, driver identification support, etc. Emergency situations can be properly tracked and alerts can be given with the incorporation of EdgeAI.

3.4.7 Summary

The edge/fog computing also supports online gaming applications, video streaming applications, industrial control applications etc. The user experience can be greatly improved in online streaming applications with edge/fog incorporation. Edge/ Fog is a promising solution for all time-critical and delay-sensitive applications.

3.5 Challenges and Research Directions

The edge/fog computing is a model where research innovations are happening at a fast pace. The impact of edge/fog integration is under study to analyse the effect of its implementation in different scenarios [2]. As the technology is still in its early stages, there are a few concerns and problems associated with it. The major challenges to be addressed with edge/fog computing and research directions are discussed below.

3.5.1 Increased Complexity

The edge/fog computing is distributed in nature with fog nodes spanning across multiple locations. The system resources also will be distributed among different locations, making the design complex. The resource allocation and maintenance should be done properly to improve the performance of the applications. Proper research initiatives have to be undertaken to reduce the complexity of distributed implementation.

3.5.2 Privacy and Security

Fog/edge computing brings processing closer to the end devices, making it more secure compared to the remote cloud. But proper authorization and authentication principles have to be designed to make the system reliable. The end devices in fog computing are connected with fog servers through a wireless network which can also result in data loss, data modification and data mishandling. Proper security policies and authorization procedures have to be designed to overcome the privacy and security challenges. Network security is always an area where research work is still ongoing. The common encryption policies and security algorithms need to be implemented and tested in the edge/fog layer to understand the potential risks.

3.5.3 Task Scheduling/Offloading

Offloading is the process of loading the given tasks to the available fog nodes in a distributed manner. The incoming requests from different devices need to be efficiently distributed to improve the performance of the application. The fog implementation should choose the appropriate load balancing algorithms to effectively offload the data. This is one of the challenges to be addressed in fog /edge implementation. Research work should be focused on designing newer algorithms to perform offloading in an efficient manner.

3.5.4 Power Consumption

The fog nodes span across multiple locations, increasing the power consumption. Not a lot of research has been done in this area. The power-constrained end devices are sometimes configured as edge devices. The edge nodes need to perform complex processing and analytics which can make the battery run down faster. This challenge has to be prioritized and appropriate solutions have to be identified.

3.5.5 Data Management

Data management is another challenge which needs to be addressed in fog implementation. The distribution of incoming data, the collection of data, data caching, encryption principles, namespace issues, etc. need to be addressed to effectively manage the data. The algorithms should be carefully designed to manage data effectively.

3.5.6 Quality of Service

The fog nodes are usually heterogeneous in nature. The QoS requirements need to be satisfied by different participating nodes. The quality has to be maintained with respect to the connectivity, data content, delivery, etc. The incorporation of different fog nodes with varying capacities can be a challenge for industrial applications [10], where QoS requirements must be guaranteed. Proper research has to be done to enable edge/fog for QoS-guaranteed applications.

3.5.7 Multi-Characteristics Fog Design System

Fog implementation should consider multiple characteristics such as bandwidth, delay, power consumption, energy efficiency, offloading, security and privacy, etc. But current research work focuses only on one or two parameters and doesn't support a combined strategy. This is an area of open research where fog designs should include multiple characteristics and an efficient system has to be designed.

The fog/edge computing suffers from a few challenges as discussed above. As the technology is still in its infant stage, there are a lot of open research problems which need to be addressed. The fog security and privacy, data offloading in distributed fog nodes, bandwidth utilization in fog, multi-characteristic fog design ideas, etc. are some of the areas where research work has to be focused. The different implementation platforms are available for fog implementation. There are numerous simulation platforms available for edge/fog implementation. The researchers can explore the features of different platforms and choose an appropriate simulator by considering the requirements. The architectural hierarchy, encryption policies, task scheduling, network usage and resource monitoring are areas where research opportunities are open.

3.6 Conclusion

Edge/fog computing has emerged as the subsidiaries to cloud computing to make processing faster and improve performance. The wide range of applications depends on the cloud for data storage, processing, performing data analytics, etc. Some applications which depend on the cloud suffer from serious latency issues. The delay and latency can be reduced by the incorporation of edge/fog computing. Edge/fog computing is implemented at the edge of the network. This greatly reduces the response time, delay, jitter and latency. Edge/fog computing has numerous benefits in terms of cost, scalability, reliability, privacy and security.

The wide range of applications currently is moving from cloud-alone platforms to an edge/fog-cloud environment. This incorporation has greatly improved the performance of the applications. Edge/fog is a promising platform which has lots of advantages over the existing cloud-alone implementation. The hybrid model which combines edge/fog and the cloud has gained wide popularity. Addressing the edge/fog-related challenges needs a thorough understanding of the concept. The different sections in this chapter will help researchers to gain knowledge about the edge/fog environment and to focus on areas which need further research inputs.

References

1. B. Alouffi, M. Hasnain, A. Alharbi, W. Alosaimi, H. Alyami, and M. Ayaz (2021) "A Systematic Literature Review on Cloud Computing Security: Threats and Mitigation Strategies". *IEEE Access*, 9: 57792–57807. doi: 10.1109/ACCESS.2021.3073203.
2. M. B. Yassein, O. Alzoubi, S. Rawasheh, F. Shatnawi, and I. Hmeidi (2020) "Features, Challenges and Issues of Fog Computing: A Comprehensive Review". *WSEAS Transactions on Computers*, 19. doi: 10.37394/23205.2020.19.12.

3. Y. Mansouri and M. A. Babar (2021) "A Review of Edge Computing: Features and Resource Virtualization". *Journal of Parallel and Distributed Comput*ing, 150: 155–183, doi: 10.1016/j.jpdc.2020.12.015.

4. M. Laroui, B. Nour, H. Moungla, M. A. Cherif, H. Afifi, and M. Guizani (2021) "Edge and Fog Computing for IoT: A Survey on Current Research Activities and Future Directions". *Computer Communications*, 180:. 210–231. doi: 10.1016/ J.COMCOM.2021.09.003.

5. J. Singh, P. Singh, and S. S. Gill (2021) "Fog Computing: A Taxonomy, Systematic Review, Current Trends and Research Challenges". *Journal of Parallel and Distributed Comput*ing, 157: 56–85. , doi: 10.1016/j.jpdc.2021.06.005.

6. H. Sabireen and V. Neelanarayanan (2021) "A Review on Fog Computing: Architecture, Fog with IoT, Algorithms and Research Challenges". *ICT Express*, 7(2): 162–176. doi: 10.1016/j.icte.2021.05.004.

7. C. H. Chen and C. T. Liu (2020) "A 3.5-Tier Container-Based Edge Computing Architecture". *Computers and Electriclal Engineering*, 93(August): 107227. , doi: 10.1016/j.compeleceng.2021.107227.

8. B. Ali, M. A. Gregory, and S. Li (2021) "Multi-Access Edge Computing Architecture, Data Security and Privacy: A Review". *IEEE Access*, 9: 18706–18721. doi: 10.1109/ ACCESS.2021.3053233.

9. Y. Siriwardhana, P. Porambage, M. Liyanage, and M. Ylianttila (2021) "A Survey on Mobile Augmented Reality with 5G Mobile Edge Computing: Architectures, Applications, and Technical Aspects". *IEEE Communications Surveys and Tutorials*, 23(2): 1160–1192. doi: 10.1109/COMST.2021.3061981.

10. I. Bouzarkouna, M. Sahnoun, N. Sghaier, D. Baudry, and C. Gout (2018) "Challenges Facing the Industrial Implementation of Fog Computing". Paper presented at 2018 IEEE 6th International Conference on the Future of the Internet of Things and the Cloud, pp. 341–348. doi: 10.1109/FiCloud.2018.00056.

Chapter 4

Reduce Overfitting and Improve Deep Learning Models' Performance in Medical Image Classification

Nidhin Raju and D. Peter Augustine

4.1 Introduction

Image classification is the process of assigning at least one name to an image, and it is one of the most important challenges in computer vision. It has several applications, including image and video recovery, video reconnaissance, web content evaluation, and biometrics. Feature coding is an important element of image classification that has received a lot of attention in recent years, with numerous coding calculations presented [1]. In general, image classification involves extracting features from images and then grouping the separated pieces. In this way, the critical question of image classification is how to extract and analyse features from images. To handle an image, traditional image classification approaches use low-level or mid-level features. Low-level features are mostly determined by grayscale thickness, variety, surface, shape, and location of data, all of which are influenced by human perception (otherwise called hand-created features). After the extraction of features in computer vision, a classifier is typically used to assign labels to various types of objects [2].

DOI: 10.1201/9781003424550-4

The deep learning technique, unlike the traditional image classification strategy, combines the process of image features extraction and grouping into a single structure. The high-level feature visualization of deep learning has proven to be superior to hand-made low-level and mid-level components, achieving excellent results in the recognition and classification of images. This is the premise of the deep learning model, which is made up of many layers, such as convolutional layers and fully connected layers that transform input data (e.g., pictures) into output data (e.g., classification outputs) while proceeding to increasingly significant level features. The primary advantage of deep learning is that it can train information-driven (or task-explicit) diverse tiered features and perform feature extraction and arrangement in a single network that is prepared from beginning to end [3, 4]. Deep learning is an effective approach in many domains, including medical image classification, due to its ability to automatically learn and extract intricate features from complex data.

Compared to the computer vision domain, applying deep learning to medical image classification has a few inherent problems. While huge databases of common, widely usable images are easily accessible and available for computer vision professionals, often for free, obtaining and using clinical images is more difficult. While massive databases of common generally beneficial images are effectively accessible and open for PC vision experts, sometimes even freely, securing and using medical images in the development of new deep learning-based innovations is a critical constraint. Despite the fact that the picture archiving and communication system (PACS) frameworks are filled with a huge number of images obtained on a regular basis in almost every clinic, medical image datasets are often small and private [3, 5]. There are two main reasons why this massive amount of stored data is not easily exploitable for clinical image analysis. The first reason is the moral and security concerns, as well as legal difficulties. Clinical information must be used according to certain rules to guarantee proper usage. To include an image in a specific report, the patient must give their informed consent, and information anonymization procedures must be followed to ensure the patient's safety. The lack of image annotations is another reason. Constructing a clinical image classification model assumes that every pixel in the picture is marked by its class, i.e., object or remaining portions, which is time-consuming and requires deeper details [6, 7].

The constraints stem not only from the imaging data but also from the nature of the relevant annotations. Obtaining specific labels for each pixel in the image is time-consuming and requires master data. As a result, professionals try to lighten the load by developing self-loading annotation tools, making insufficient annotations, or publicly promoting non-master labels. Managing label noise when creating classification models is a test in any of these scenarios. The nature of the outcome comments, which can be treated as fuzzy, is often substandard, with variables such as information, the objective of the images, visual discernment, and weakness playing a significant role. As a result, developing a deep learning framework based on such data necessitates careful consideration of the most effective way of managing noise and exposure in the real world, which is still an ongoing topic of research [8].

As a result, properly progressing from limited clarified data is an important topic of investigation that designers and specialists must focus on. There are commonly different strategies to increase the dataset size as the deep learning-based classification approach. Data augmentation is one of the strategies which refers to enlarging the dataset by adding extra images, either through simple activities such as interpretations and revolutions or through more advanced methods. In many applications, instead of traditional photos, 3D medical volumes are replaced with heaps of autonomous 2D images to provide additional information in the model [9]. Sub-volumes or image patches are frequently eliminated from images to increase the amount of data available. The undeniable flaw is that the anatomic setting in headings symmetrical to the slice plane is completely eliminated. In light of Generative Adversarial Networks (GAN), another technique to cope with making fabricated images to enlarge the dataset has just been offered. The GANs are a unique type of neural network model in which two models are prepared simultaneously: one is the generator, which is focused on creating manufactured images from commotion vectors, and the other is the discriminator, which is focused on distinguishing between genuine and fake images produced by the generator. GANs are created by addressing an advancement issue that the discriminator tries to improve and the generator tries to reduce [10, 11].

4.2 Deep Convolutional Neural Networks (DCNNs)

DCNNs are complex neural networks made up of different levels of transformations known as "layers." The majority of DCNNs can be divided into two sections. A feature extraction part extracts various features from inputs. The other part is the classification, which actually handles the given issue by using the feature characterization removed by the element extraction component. The multi-layer perceptron, which is a typical classifier in neural networks, is commonly accepted as a classification component. The advantage of DCNNs is the exceptional ability of the feature extraction part to acquire a decent contracted portrayal of information sources appropriate for a given task [12, 13]. Furthermore, back-propagations models enable DCNNs to acquire such good portrayals, which have been handcrafted by experts in the way of traditional image recognition, with the start-to-finish streamlining alongside the classification section. DCNNs' feature extraction layers primarily consist of "convolution" and "spatial pooling" layers. Convolution layers function as nearby element extractors, similar to neocognitron cells. This layer maps inputs into other activation maps, which accentuate various nearby elements with convolution activity in a manner similar to image processing. Spatial pooling layers spatially pack the actuation guide in order to tolerate minor nearby distortions and reduce the dimensionality of the activation map. This capability is similar to that of C-cells and contributes to the abstraction of various leveled features in our visual data handling [14, 15]. Convolutional neural networks (CNNs) are classified as deep learning networks and are commonly used to analyse and classify images. They are made up of an input and an output layer, with several hidden layers in between. The hidden layers are made up of a few convolutional layers that extract the various features of the images.

A CNN is represented by the following non-linear function:

$$c_i = C(M; \theta) \tag{4.1}$$

which transforms an image $M \in R^{W \times W}$ of size W×W to a vector $c_i = \left(c_1, c_2, \dots c_n \right)^T$, where ci [0,1] and signifies the chance that the Image M belongs to one of the n classes: $i = 1 \dots n$. The number of K parameters used to map the input image to the vector c_i is $\theta = \left\{ \theta_1, \theta_2, \dots \theta_K \right\}$.

CNN training can be viewed as a non-linear optimization problem:

$$\hat{\theta} = arg_{\theta,} min\, K\{k^{(1)}, k^{(2)} \dots, k^{(N)}\}^{(\theta)} \tag{4.2}$$

N is the number of images used to train the CNN in this case, and

$$K\{k^{(1)}, k^{(2)} \dots, k^{(N)}\}^{(\theta)} = \frac{1}{N} \sum_{j=1}^{N} D_{j,} vE^{(i)T}\, logC\left(I^{(i)}; \theta \right) \tag{4.3}$$

$$D_i = \frac{\dfrac{1}{O_i}}{\sum_{l=1}^{n}\dfrac{1}{O_i}} \tag{4.4}$$

$$\theta_{\lambda+1} = \theta_\lambda - \alpha \Delta L\left(\theta_\lambda \right) + \gamma \left(\theta_l - \theta_{\lambda-1} \right) \tag{4.5}$$

4.3 Medical Image Classification Using Transfer Learning

Transfer learning is one of the techniques of deep learning in one task to improve speculation in another. For example, given the extensive measure of structure information provided in task-1, a model learning the visual highlights of structures will actually want to learn and summarize the visual elements of vehicles in task-2. The feature extraction and fine-tuning are the two types of transfer learning in CNN [16]. Feature extraction is performed in the previously mentioned transfer learning example; when a CNN gains features from task-1, then uses a similar base convolutional and pooling layers of networks by supplanting the completely associated layer with the task-2 explicit classifier. The loads weights of the convolutional base will be frozen during this activity, and only the new classifier is ready for vehicle classification [17, 18]. The features gained from the convolutional base are reusable, but characterizations learned in the completely associated layers are intended for the classes on which the model is trained. The weights of the convolutional base are slightly adjusted to match the task-2 issue after preparing the network on structures information in fine-tuning. This necessitates the unfreezing of a few high-level layers of the convolutional base and the addition of another classifier, along with its learned parameters while the whole convolutional base is frozen unlike the feature extraction. When performing this type of transfer learning, the unfrozen layers of the CNN are prepared alongside the completely associated layer when new data is processed. Multi-stage transfer learning employs either extraction of features or fine-tuning at least once or twice, enabling tweaked training pipelines and higher adaptability of reusing characterizations gained from previous stages in the resulting preparatory activities [19, 20].

4.3.1 Challenges When Using Transfer Learning in Medical Image Classifications

A significant portion of the medical image classification focuses on, in light of transfer learning, use of variations of exemplary CNN models, for example, Mobilenet, Densenet, Inception, VGG19. In medical image classifications, CNNs are over-parameterized to improve execution. The standard deep CNNs learn more slowly than lightweight models during the training interaction due to over-parameterization. This challenge may result in increased training times and longer epochs as a result of grossly insufficient learning [21]. For example, a Resnet is combined with the Random Forest algorithm for brain hemorrhage characterization tasks, which demonstrated a superior level of exactness compared to the standard Resnet model, despite the fact that it required a greater number of epochs and a longer training time to achieve good accuracy. The transfer learning offers significant advantages over using the model without any preparation; however, model sizes and parameters have negative effects on account of calibrating ImageNet pre-trained models. Recent research includes additional consideration modules for acquiring discriminatory elements from the deep layers and rejecting unnecessary inclusions. Using these modules demonstrated the ability to improve precision while increasing the computational weight [22, 23].

Generally, shorter training times and lower computational costs would be advantageous to the medical image classification field if model viability were not compromised. Standard CNN models are effective in medical image classification due to their high level of depth and large number of layers, but they incur additional computational costs. The Inception model has 7 million parameters, which is fewer than other excellent CNNs but too expensive to be trained on a standard computer device [24]. While there has been significant development in Central Processing Units and Graphics Processing Units recently, it is not practical in some real-world applications. As this is still in its early stages, this current situation provides an opportunity to investigate lightweight architectures (LWAs) that can be used in cell phones to improve convenience and productivity. A dataset is used labeled if it contains appropriate image comments for each of the available classes. For example, if a cardiac image dataset has two classes, "cardiac" and "usual," each image should be clarified with both of these classes [25].

Labeled datasets are important for medical image classification because deep learning models rely on a large number of labeled images for training in order to provide productive image classification. Collecting a large number of labeled clinical image datasets can be a difficult task, in contrast to regular images. With access to decent labeled image data, transfer learning can also succeed and make more powerful arrangements. Traditional transfer learning techniques perform well in the absence of information, but there is still room for improvement [26]. In normal images, the differences in features are critical from one image to the next, with the images having a variety of lighting and shapes, yet in medical images, these distinctions can be minute, depending on the problem, making it difficult to learn characterizations with limited information while using transfer learning strategy. multi-stage transfer learning is one viable strategy for overcoming this challenge.

4.3.2 Applications of Transfer Learning in Medical Image Classifications

While traditional transfer learning employs fine-tuning and includes extraction of features, multi-stage transfer learning employs it at least several times. This allows models to gain features from a related domain (medical) and perform better on track due to feature similarities. Modality bridge transfer learning was possibly the first assessment to perform two-stage transfer learning in medical image classification, using images from MRIs, CTs, and X-rays as scaffold datasets and objective information bases from similar areas with less-commented-on images, whereas the source has been ImageNet. The projection function from the input dataset into the element space of the scaffold dataset is learned; then, non-direct planning of component space from the source to connect is performed; and finally, the classifier learns based on the projection task being used by the span dataset [27, 28]. This process maintains the area variation of the issue dataset and the source dataset. A VGG16 model can be used to perform the analyses, which results in working on the precision in each of the three modalities, such as MRI, CT, and X-ray [29]. Lightweight architectures have fewer constraints and are smaller in size than standard deep CNNs, yet they appear to produce comparable accuracies. LWAs allow us to prepare these models quickly and efficiently. MobileNetsV2, NasNets, EfficientNets are examples of LWAs that are opening the way for more proficient image classification products. Each of these models has unique compositional advantages that result in improved image recognition tasks. MobileNets, for example, uses depth-wise distinct convolutions, whereas SqueezeNet achieves better exactness by replacing 3×3 convolutional channels with 1×1 channels to decrease computational expenses during basic training. The use of LWAs in medical image classification is still in its early stages, but the field is rapidly progressing with recommendations of new ways to further develop proficiency, as mentioned in Section 4.3.1. A large number of these models are used for transfer learning in medical image classification. Due to its engineering of fewer boundaries, the SqueezeNet model outperformed all other models in terms of speed. Despite the fact that SqueezeNet achieved excellent results with additional complex models, there were some significant misclassifications of precancerous images as harmless, indicating that there are issues in LWAs and that further analysis is necessary [30, 31].

4.4 Overfitting Prevention Techniques

Overfitting occurs when your model achieves a strong match on the preparation data but fails to perform well on new, hidden data. Typically, the model learns well-defined designs for training data that are superfluous in other data. Overfitting can be identified by looking at approval measurements, which are similar to accuracy or loss. Normally, the validation metric stops working after a certain number of epochs and begins to decline later. The training metric continues to improve as the model aims for the best fit for the training data. When a model tries to predict a pattern in data that is too noisy, it is called overfitting. This is due to a model that is overly sophisticated and has a huge number of boundaries. In view of the fact that the pattern does not represent the truth existing in the data, an overfitted model is incorrect. This can be determined if the model produces excellent results on the visible data (training set) but fails miserably on the hidden data (test set). The deep learning model's goal is to generalize in the best possible way everything from the training data to any information from the area of the problem. This is critical since we consider our model should generate predictions based on data it has never seen before [32].

In deep learning, the main criterion is that we should be able to perform effectively on new, previously unknown inputs in addition to those for which our model was trained. The ability to perform effectively on previously unknown data sources is referred to as generalization. A deep learning network's goal is to produce a final model that performs well on both the data we used to train it and the new data on which the model can be used to make forecasts. A model that learns the training dataset reasonably well and performs well on the hold-out dataset is a good fit model. It's a typical pitfall in deep learning when a model tries to fit all of the trained data and ends up retaining the patterns as well as the noise and irregular variances. Because of invisible data conditions, these models fail to summarize and perform adequately, negating the point of the model. The high chance of the model exhibition indicates an overfitting problem. The model's training duration or architecture depth may cause it to overfit [33]. If the model trains for an extremely long time on the training data or is overly complex, it learns the noise or superfluous data from the dataset. A model that fails to adequately learn the problem, performs poorly on a training dataset, and does not perform well on a hold-out test is an underfit model. A model fit can be thought of as a predisposition difference compromise. Underfit models have a strong bias but a limited variance. It cannot become acquainted with the issue, regardless of the specific examples in the training data. An overfit model has a limited bias and a strong variance. The model learns the training data extremely well, and its execution differs significantly from that of new unseen data or even quantifiable noise added to samples in the training dataset [34]. The easy ways to reduce overfitting are:

■ by adding more data
■ by changing the complexity of the model.

The advantage of extremely deep neural networks is that their performance improves as they handle increasingly large datasets. A model with a near-limitless number of data will eventually level out as far as what the network's limit is fit for learning. A model can overfit a training dataset if it has the necessary capability. Reduce the model's limit to reduce the likelihood of the model overfitting the preparation dataset to the point where it no longer overfits. The complexity of a brain network model is defined by both its construction in terms of nodes and layers and its boundaries in terms of weights. In this way, we can change the complexity of a model to reduce overfitting in two different ways:

■ Change the number of weights by modifying the network structure.
■ Change the value of weights by changing the parameters.

In neural networks, the complexity can be changed by adjusting the number of versatile parameters. Another primary method for controlling the complexity of a model is through the use of regularization, which includes the inclusion of a penalty term to the error method. Massive weights will almost always result in sharp advances in activation functions and, as a result, enormous changes in output for small changes in inputs. It is more reasonable to concentrate on techniques that compel the size of the weights of a neural network, because a single model structure can be characterized that is under-compelled, for example, it has a much higher limit than is expected for the issue, and regularization can be used during training to ensure that the model doesn't overfit. In such cases, execution may be improved because the additional limit can be oriented on better learning generalizable ideas in the work [35, 36]. Regularization strategies are those that attempt to decrease overfitting by keeping network weights to a minimum. Regularization relates to a class of approaches that add extra data to change a poorly presented problem into a more consistent, well-presented matter. When small changes in the given data cause large changes in the arrangement, a problem can be said to be poorly presented. Because of the uncertainty surrounding the information, arrangements are untruthful because minor estimation errors in constraints can be extraordinarily amplified and lead to absurdly diverse reactions. The concept behind regularization is to utilize useful data to repeat a poorly presented issue in a consistent structure. Regularization strategies are so extensively used to reduce overfitting that the term regularization could be used to refer to a strategy that improves the generalization error of a model. Regularization is any adjustment to a learning model that is intended to reduce its generalization error but not yet its training faults. Regularization is one of the primary concerns in the field of deep learning, challenged in importance only by optimizing [37, 38].

4.4.1 Weight Regularization

The easiest and possibly most common regularization strategy is to add a penalty to the loss measurement in relation to the size of the loads in the model. Neural networks become familiar with a variety of loads that best guide contributions to yields. A network with massive network weights may be an indication of an unstable architecture in which minor changes in input can result in significant changes in the outcome. This could indicate that the model overfits the training dataset and will probably perform poorly when forecasting on new data. One way to solve this is to update the learning model to empower the networks to keep the weights light. This is known as weight regularization, and it is commonly used as an overall strategy to reduce overfitting of the training dataset and work on the model's generalization. While training a neural network model, we should become acquainted with the network's weights by using stochastic gradient descent and the training dataset. The more we train the model, the more specific the weights will become to the data, causing the training data to be overfit. The weights will fill in size to deal with the samples available in the training data. Massive weights render the model unstable. When the weights are specific to the training dataset, minor variations or factual noise on the standard sources of the data will cause huge differences in the results. A network with huge weights is more complex than one with smaller weights. It indicates a model that is overly concerned with training data. Generally, we recommend using simpler models to solve problems and models with lighter weights. Another possibility is that there are numerous input factors, each with varying degrees of relevance to the outcome variable. We can use strategies to help us choose input factors at times, but the interrelationships between factors aren't always obvious. Having few or no weights for less important or superfluous contributions to the network will permits the model to grow and learn. This will also contribute to a simpler model [39, 40].

Deep learning models are capable of effortlessly developing a rich internal representation from basic data. Feature representation learning is the term used for this. Better-informed representations might therefore encourage the entry of better knowledge into the domain, for instance through the representation of learned features as well as better prophetic models that use the learned highlights. The fact that learned features may be overly particular to the training data or overfit and not perform well with other models is a problem. Massive characteristics in the learned representation may be a sign that it is overfitted. Activity regularization offers a method to enable learned representations, which are the outcome or commencement of the model's hidden layer, to stay insufficient and underdeveloped. Feature learning can be performed by advanced deep learning models, i.e., during network training, the model will separate the notable highlights from the information designs or learn highlights. These features could be used in the model to forecast a relapse amount or to predict a class as an incentive for characterization. These inner portrayals are significant. The result of a hidden layer within the model is addressing the model's learned features by then in the model [39]. Bigger weights lead to harsher penalties as well as worse loss scores. The model will therefore be forced to have more moderate weights, such as weights that are no larger than what is necessary for it to perform effectively on the training dataset. We refer to this penalty as weight regularization since smaller weights are seen to be more typical or less specialized. When this method of penalizing model coefficients is applied to other ML models, such as logistic or linear regression, it is quite possible that shrinkage will be used to refer to it because the punishment causes the coefficients to contract during the optimization system [40].

Overfitting is indicated by high weights in a neural network. A massive model has most likely taken in the measurable noise in the training data. As a result, the model is unstable and extremely sensitive to changes in the information factors. As a result, the overfit network performs poorly when generating predictions based on fresh discrete data. A well-known and appealing approach to resolving the issue is to refresh the loss function that is refined throughout training to consider the weights' size. This is characterized as a penalty, since the more the model's weights grow, the more the model is punished, resulting in greater disadvantage and, therefore, greater upgrades. As a result, the penalty encourages weights to be little or no larger than planned during the training period, reducing overfitting. A disadvantage of using a penalty is that, while it encourages the model to take on smaller weights, it does not compel smaller weights. In any event, a neural network equipped with a weight regularization penalty may enable massive weights, at times quite enormous weights [41, 42].

The learning model can be upgraded to assist the network in utilizing light weights. One way to accomplish this is to modify the loss measurement used in the network's advancement to take into account the size of the weights. Keep in mind that when we train a model, we limit a loss measurement. We can add the current size of all loads in the model or include a layer to this estimation when calculating the difference between the anticipated and expected values in a batch. This is referred to as a penalty because we are punishing the model in relation to the size of the weights in the model. Larger weights result in a harsher penalty, as well as a higher loss score. The optimization algorithm will then push the model to have more modest weights, such as weights no larger than what is required to perform well on the training dataset. Tinier weights are regarded as more normal or less abnormal, use this penalty as weight regularization in this way [39]. Higher weights result in a harsher punishment, as well as a higher loss score. The advancement calculation will then drive the model to have more modest weights, such as weights no larger than what is necessary to perform well on the training dataset. We refer to this penalty as weight regularization since lighter weights are considered more common or less particular. When this technique of punishing model coefficients is used in various AI models, it might be addressed as shrinkage, because the punishment energizes the coefficients to retreat during the progression interaction.

4.4.2 Activity Regularization

Deep learning models can facilitate feature learning. The model will naturally extract the notable features from the input data or learn features during network training. These features could be used in the model to forecast a relapse amount or anticipate a class as an incentive for classification. These interior characterizations are specific things. The features learned by the model at that point in the network are represented by the output of a hidden layer within the network. There is an area of research focused on the productive and successful programmed learning of features, which is frequently examined by having a network reduce a contribution to a minor scholarly feature before using a second network to remake the exact input from the learned feature. These models are known as auto-encoders or encoder-decoders, and their learned elements can be helpful in learning more about the domain and in forecasting analytics. The learned elements, or encoded inputs, should be large enough to capture the salient characteristics of the data while also being narrow enough not to overfit the specific models in the training dataset. As a result, there is conflict between the expressiveness and speculation of the learned features [43].

The network's loss function can be refreshed to punish models in relation to the magnitude of their actuation. This is similar to weight regularization in that the loss function is refreshed to punish the model based on the magnitude of the loads. This type of punishment or regularization is known as initiation regularization or action regularization because the result of a layer is referred to as its activation or action. The result of an encoder or, for the most part, the result of a hidden layer in a neural network could then be viewed as the model's characterization of the issue.

As a result, this type of penalty is also known as representation regularization. The desire to have few initiations or even few enactments with generally zero qualities is also known as a desire for sparsity. As a result, this type of penalty is also known as sparse feature learning. Sparsity is typically sought when a bigger than required hidden layer is used to learn features that may enable overfitting. The presentation of a sparsity punishment mitigates this problem and encourages better speculation. A scanty overcomplete learned feature has been shown to be more successful than other types of learned features, offering better vigor to commotion and even changes in the data, for example, learned features of images may have further developed invariance to the position of obstacles in the image [44, 45].

An activation penalty can be applied for each layer, even if only at one layer that is the focal point of the learned characterization, such as the result of the encoder model or the center of an auto-encoder model. A constraint that adds a penalty corresponding to the magnitude of the layer's vector result can be applied. Because the initiation values may be positive or negative, we cannot simply aggregate the qualities. There are two common methods for determining the size of the activation as follows:

■ Amount of the outright actuation values, also known as the L1 norm.
■ The L2 norm is the sum of the squared actuation values.

The L1 norm promotes sparsity by, for example, allowing a few enactments to become zero, whereas the L2 norm promotes small initiation values overall. The use of the L1 norm may be a more commonly involved penalty for initiation regularization. A hyperparameter indicating the sum or degree to which the misfortune capability will weight or suffer in light of the consequence should be determined. Normal qualities are measured on a logarithmic scale between 0 and 0.1. Activity regularization could be applied in conjunction with other regularization procedures, such as weight regularization. A common approach is activation regularization. It may be used with most, if not all, types of neural network models, including the best-known network types, such as Multilayer Perceptrons, Convolutional Neural Networks, etc. Activity regularization could be the best option for model types that are looking for an excellent learnt depiction. These contain auto-encoder models and encoder-decoder models. The L1 standard, which enables sparsity, is the most commonly recognized activation regularization. Investigate several routes for different types of regularization, such as the L2 standard or using both the L1 and L2 norms concurrently. The rectified linear activation function, also known as relu, is an actuation capability that is currently widely used in the hidden layer of deep neural networks. Despite classic initiation works such as tanh and sigmoid (strategic capability), the relu capability allows for definite zero characteristics without issue [45]. This makes it a viable contender when learning limited representations, for example, with the L1 vector norm for regularization. It is usual to provide minor characteristics for the regularization hyperparameter, which affects the commitment of each enactment to the penalty. Start by experimenting with values on a log scale, such as 0.1, 0.001, and 0.0001. Then, at that point, do a grid search at the crucial degree that demonstrates the most dedication. Rescaling input factors to have the same scale is generally a good approach. When input factors have different scales, the magnitude of the organization's burdens will fluctuate correspondingly. Massive loads can drown the nonlinear exchange capacity and reduce the change in the layer's outcome. This might be a problem when using activation regularization. This problem can be solved by standardizing or normalizing input components. Design the layer chosen to be the trained features, such as the encoder result or the bottleneck in the auto-encoder, to include more nodes than may be necessary. This is known as an overcomplete representation, and it will encourage the model to overfit the preparatory samples. This can be offset by regions of strength for a regularization to stimulate a rich learnt depiction that is also insufficient [44, 45].

4.4.3 Adding Dropout Layers

Deep learning neural networks are likely to rapidly overfit a training dataset with few models. Ensemble networks with various model setups are known to reduce overfitting; however, this requires the additional computational cost of training and maintaining different models. By irregularly exiting nodes during training, a single model can be used to recreate having an immense number of different model designs. This is known as dropout and, all else being equal, offers a very computationally modest and incredibly compelling regularization strategy to reduce overfitting and generalization errors in deep learning models. Massive neural networks built on relatively small datasets have the potential to overfit the training data. When the model is evaluated on unseen data, such as a test dataset, it performs poorly since it has learned the actual noise in the training data. Overfitting leads to speculation error. One strategy for reducing overfitting is to fit all the different neural networks to the same dataset and average the predictions from each model. This isn't achievable right away, but it may be roughly estimated using a small group of different models, or gathered ones [46, 47].

Dropout is a regularization technique that simulates the equal training of a huge number of different models at a time. Throughout the training, a few node outputs are accidentally neglected or eliminated. As a result, the layer seems to be and is treated as a layer with a different set of nodes and network than the preceding layer. In effect, every layer update during training is carried out from a different angle than the configured layer. Dropout makes the learning process noisy, forcing nodes in one layer to probabilistically assume full responsibility for the information sources. This perspective suggests that perhaps dropout distinguishes situations when network layers cooperate to correct errors from prior layers, strengthening the model. Dropout replicates a sparse activation from a particular layer, which oddly prompts the model to adjust to a sparse depiction as a matter of fact as a side impact. As a result, it might be used as a replacement for activity regularization to enable sparse depictions in auto-encoder models. Dropout occurs per layer in a neural network model. The majority of layer types, including large totally associated layers, convolutional layers, and recurrent layers prefer to be used with it. Dropout may be applied to any or all model hidden levels as well as the input or visible layer. The output layer does not make use of it [48, 49].

4.4.4 Noise Regularization

Adding random noise is one technique to address the growing conjecture blunder and work on the mapping issue's structure. The increase in noise during the development of a neural network model affects its regularization and, consequently, its performance. It has been demonstrated that weight regularization approaches have a similar effect on the capacity for misfortune as the extension of a penalty period. The size of the training dataset is essentially increased by introducing noise. The input variables are altered randomly each time a training sample is introduced into the model, making each exposure to the model unique. This makes the addition of noise to input samples a straightforward method of data augmentation. The addition of noise reduces the network's ability to recall training samples since they change all the time, leading in smaller network weights and a more resilient network with reduced generalization error. The noise implies that fresh samples are being pulled from the domain around known samples, smoothing the structure of the input space. Because of this smoothing, the mapping function may be easier for the network to learn, resulting in better and faster learning [50, 51].

4.4.5 Stop Training with the Early Stopping Method

When preparing a large model, there will come a moment when the model will stop executing and begin learning the factual noise in the training dataset. This overfitting of the training dataset will result in a rise in generalization error, making the model less useful in forecasting unseen data. The difficulty is to prepare the model to acquire the planning from contributions to yields while not training the model for so long that it overfits the training data. One solution is to use the number of training epochs as a hyperparameter and train the model numerous times with different values before selecting the number of epochs that result in the greatest performance on the training or holdout test dataset. The disadvantage of this strategy is that it necessitates the training and discarding of several models. This can be computationally inefficient and time-consuming, especially for big models trained over days or weeks on enormous datasets [52, 53]. Another option is to train the model only once for a large number of training epochs. After each epoch of training, the model is assessed on a holdout validation dataset. If the model's performance on the validation dataset begins to deteriorate, the training process is terminated. The model that is used when training is stopped is known to have strong generalization performance. This is known as early stopping, and it is one of the oldest and most extensively used types of neural network regularizations [54, 55].

4.5 Conclusion

One of the important components for medical image assessment is deep learning. It has been used successfully in target identification, segmentation and classification. The advance of deep learning in the clinical sector is dependent on the gathering of vast amounts of clinical information, which has multi-modular features and thus provides a lot of rich information for deep learning. In terms of disease treatment, deep learning may not only locate the injured spot, but also distinguish and identify specific lesions. When the injury is true, several identifying models can also fragment the injured area. While deep learning has many advantages, it also has some drawbacks. The deep learning approach is heavily reliant on informational collections. Every deep learning network necessitates massive information for training, making data index acquisition really difficult. The fundamental reason is that the pixel highlights in the first information image are overly intricate, thus a future improvement pattern is to focus on developing a model with a smaller data size. Deep learning may evolve so quickly in the medical industry because it is inextricably linked to a wide range of treatment procedures. The goal of adequately applying deep learning to all phases of medical care is becoming more complicated. It is determined by two factors: the perpetual iteration of technology and the continuing accumulation of medical experience. On the other hand, such deep learning models tend to overfit issues mainly because of the lack of available data in the training set and network architecture. We present some solutions to the overfitting issues and leverage the performance of the deep learning models in this chapter.

References

1. C. Affonso, A. L. D. Rossi, F. H. A. Vieira, and A. C. P. de Carvalho (2017) "Deep Learning for Biological Image Classification." *Expert Systems with Applications*, 85: 114–122. doi:10.1016/J.ESWA.2017.05.039.
2. S. R. M. Zeebaree et al. (2020) "Deep Learning Models Based on Image Classification: A Review of Facial Expression Recognition Based on Hybrid Feature Extraction Techniques," *International Journal of Science and Business*, 4(11): 75–81. doi:10.5281/zenodo.4108433.
3. S. Suganyadevi, V. Seethalakshmi, and K. Balasamy (2021) "A Review on Deep Learning in Medical Image Analysis." *International Journal of Multimedia Information Retrieval*, 11(1): 19–38. doi:10.1007/S13735-021-00218-1.

4. B. Kantheti and M. K. Javvaji (2022) "Medical Image Classification for Disease Prediction with the Aid of Deep Learning Approaches." Paper presented at 6th International Conference on Intelligent Computing and Control Systems (ICICCS), June, pp. 1442–1445. doi:10.1109/ICICCS53718.2022.9788144.

5. Y. Li (2022) "Research and Application of Deep Learning in Image Recognition." Paper presented at 2022 IEEE 2nd International Conference on Power, Electronics and Computer Applications (ICPECA), Shenyang, China, March 2022, pp. 994–999. doi:10.1109/ICPECA53709.2022.9718847.

6. M. Tsuneki (2022) "Deep Learning Models in Medical Image Analysis." *Journal of Oral Biosciences*, 64(3): 312–320.,doi:10.1016/J.JOB.2022.03.003.

7. N. Raju (2017) "Text Extraction from Video Images." *International Journal of Applied Engineering Research*, 12: 14750–14754 (accessed August 12, 2022). Online. Available at: www.ripublication.com

8. A. Rehman, M. Ahmed Butt, and M. Zaman (2021) "A Survey of Medical Image Analysis Using Deep Learning Approaches." In *Proceedings of 5th International Conference on Computing Methodologies and Communication (ICCMC)*, April, pp. 1334–1342. doi:10.1109/ICCMC51019.2021.9418385.

9. M. A. Abdou (2022) "Literature Review: Efficient Deep Neural Networks Techniques for Medical Image Analysis." *Neural Computing and Applications*, 34(8): 5791–5812.,doi:10.1007/S00521-022-06960-9/TABLES/5.

10. S. S. Yadav and S. M. Jadhav (2019) "Deep Convolutional Neural Network Based Medical Image Classification for Disease Diagnosis."*Journal of Big Data*, 6(1): 1–18. doi:10.1186/S40537-019-0276-2/TABLES/16.

11. M. Frid-Adar, I. Diamant, E. Klang, M. Amitai, J. Goldberger, and H. Greenspan, (2018) "GAN-Based Synthetic Medical Image Augmentation for Increased CNN Performance in Liver Lesion Classification." *Neurocomputing*, 32(1): 321–331., doi:10.1016/J.NEUCOM.2018.09.013.

12. P. Malhotra, S. Gupta, D. Koundal, A. Zaguia, and W. Enbeyle (2022) "Deep Neural Networks for Medical Image Segmentation," *Journal of Healthcare Engineering*. Online. doi:10.1155/2022/9580991.

13. X. Wang et al. (2022) "Detection and Classification of Mandibular Fracture on CT Scan Using Deep Convolutional Neural Network,." *Clinical Oral Investigation*, 26(6): 4593–4601. doi:10.1007/S00784-022-04427-8.

14. X. Zhang, V. C. Lee, J. Rong, J. C. Lee, and F. Liu (2022) "Deep Convolutional Neural Networks in Thyroid Disease Detection: A Multi-Classification Comparison by Ultrasonography nd Computed Tomography." *Computer Methods and Programs in Biomedicine*, 220: 106823.,doi:10.1016/J.CMPB.2022.106823.

15. L. Wang et al.(2022) "Trends in the Application of Deep Learning Networks in Medical Image Analysis: Evolution between 2012 and 2020." *European Journal of Radiology*, 146: 110069.,doi:10.1016/J.EJRAD.2021.110069.

16. X. Yu, J. Wang, Q. Q. Hong, R. Teku, S. H. Wang, and Y. D. Zhang (2022) "Transfer Learning for Medical Images Analyses: A Survey." *Neurocomputing*, 489: 230–254.,doi:10.1016/J.NEUCOM.2021.08.159.

17. J. Meng, Z. Tan, Y. Yu, P. Wang, and S. Liu (2022) "TL-Med: A Two-Stage Transfer Learning Recognition Model for Medical Images of COVID-1." *Biocybernetics and Biomedical. Engineering*, 42(3): 842–855. doi:10.1016/J.BBE.2022.04.005.

18. L. Alzubaidi et al. (2021) "Novel Transfer Learning Approach for Medical Imaging with Limited Labeled Data." *Cancers*, 13(7):1590, doi:10.3390/CANCERS13071590.

19. J. Wang, H. Zhu, S. H. Wang, and Y. D. Zhang (2021) "A Review of Deep Learning on Medical Image Analysis." *Mobile Networks and Applications*, 26(1): 351–380.,doi:10.1007/S11036-020-01672-7/TABLES/11.

20. S. Niu, M. Liu, Y. Liu, J. Wang, and H. Song (2021) "Distant Domain Transfer Learning for Medical Imaging." *IEEE Journal of Biomedical and Health. Informatics*, 25(10): 3784–3793. doi:10.1109/JBHI.2021.3051470.

21. G. Ayana, K. Dese, and S. W. Choe (2021,) "Transfer Learning in Breast Cancer Diagnoses via Ultrasound Imaging." *Cancers*, 13(4): 738.doi:10.3390/CANCERS13040738.

22. F. Altaf, S. M. S. Islam, and N. K. Janjua (2021) "A Novel Augmented Deep Transfer Learning for Classification of COVID-19 and Other Thoracic Diseases from X-Rays," *Neural Computing and Applications*, 33(20): 14037–14048.,doi:10.1007/S00521-021-06044-0/TABLES/7.

23. M. A. Morid, A. Borjali, and G. Del Fiol (2021) "A Scoping Review of Transfer Learning Research on Medical Image Analysis Using ImageNet." *Computers in Biology and Medicine*, 128: 104115. doi:10.1016/J.COMPBIOMED.2020.104115.

24. D. R. Sarvamangala and R. V. Kulkarni (2022) "Convolutional Neural Networks in Medical Image Understanding: A Survey." *Evolutionary Intelligence*, 15(1): 1–22. doi:10.1007/S12065-020-00540-3/FIGURES/2.

25. M. Arbane, R. Benlamri, Y. Brik, and M. Djerioui (2021) "Transfer Learning for Automatic Brain Tumor Classification Using MRI Images." Paper presented at 2nd International Workshop on Human-Centric Smart Environments for Health and Well-Being, IHSH 2020, pp. 210–214, February. doi:10.1109/IHSH51661.2021.9378739.

26. S. Cyriac, N. Raju, and S. Ramaswamy (2021) "Comparison of Full Training and Transfer Learning in Deep Learning for Image Classification." *Lecture Notes on Networks Systems*, 290: 58–67. doi:10.1007/978-981-16-4486-3_6/COVER.

27. D. Karimi, S. K. Warfield, and A. Gholipour (2021) "Transfer Learning in Medical Image Segmentation: New Insights from Analysis of the Dynamics of Model Parameters and Learned Representations." *Artificial Intelligence in Medicine*, 116: 102078. doi:10.1016/J.ARTMED.2021.102078.

28. M. O. Aftab, M. Javed Awan, S. Khalid, R. Javed, and H. Shabir (2021) "Executing Spark Big DL for Leukemia Detection from Microscopic Images Using Transfer Learning." Paper presented at 2021 1st International Conference on Artificial Intelligence and Data Analysis (CAIDA 2021), April, pp. 216–220. doi:10.1109/CAIDA51941.2021.9425264.

29. N. Raju, Anita H. B., and P. Augustine (2020) "Identification of Interstitial Lung Diseases Using Deep Learning." *International Journal of Electrical and Computer Engineering*, 10(6): 6283–6291. doi:10.11591/IJECE.V10I6.PP6283-6291.

30. L. Gaur, U. Bhatia, N. Z. Jhanjhi, G. Muhammad, and M. Masud (2021) "Medical Image-Based Detection of COVID-19 Using Deep Convolution Neural Networks." *Multimedia Systems*, 1: 1–10. doi:10.1007/S00530-021-00794-6/TABLES/6.

31. J. M. Valverde et al. (2021) "Transfer Learning in Magnetic Resonance Brain Imaging: A Systematic Review." *Journal of Imaging*, 7(4): 66. doi:10.3390/JIMAGING7040066.

32. M. M. Bejani and M. Ghatee (2019) "Regularized Deep Networks in Intelligent Transportation Systems: A Taxonomy and a Case Study." *Artificial Intelligence Review*, 54(8): 6391–6438. doi:10.1007/s10462-021-09975-1.

33. J. M. Kernbach and V. E. Staartjes (2022) "Foundations of Machine Learning-Based Clinical Prediction Modeling: Part II—Generalization and Overfitting." *Acta Neurochirurgica Supplement*, 134: 15–21. doi:10.1007/978-3-030-85292-4_3/COVER.

34. C. F. G. dos Santos and J. P. Papa (2021) "Avoiding Overfitting: A Survey on Regularization Methods for Convolutional Neural Networks." *ACM Computing Surveys*, May. doi:10.1145/3510413.

35. C. K. Chan and G. T. Chala (2022) "Overfit Prevention in Human Motion Data by Artificial Neural Network." *Platform: A Journal of Engineering*, 5(1): 29–37. Available at: https://myjms.mohe.gov.my/index.php/paje/article/view/12982 (accessed August 12, 2022).

36. M. K. Rusia and D. K. Singh (2021) "An Efficient CNN Approach for Facial Expression Recognition with Some Measures of Overfitting." *International Journal of Information Technology*, 13(6): 2419–2430. doi:10.1007/S41870-021-00803-X.

37. B. El Asri, M. Rhanoui, and B. Sabiri (2022) "Mechanism of Overfitting Avoidance Techniques for Training Deep Neural Networks Building Scalable Cloud Computing Based on Variant-Rich Service Component Architecture." In *Proceedings of the 24th International Conference on Enterprise Information Systems (ICEIS 2022)*. doi:10.5220/0011114900003179.

38. Z. Biczo, S. Szenasi, and I. Felde (2022) "Safe Overfitting of Boosted Tree Algorithm in Heat Transfer Modeling." Paper presented at SAMI 2022–IEEE 20th Jubilee World Symposium on Applied Machine Intelligence Informatics, pp. 379–382. doi:10.1109/SAMI54271.2022.9780808.

39. C. Ou et al. (2022) "Quality-Driven Regularization for Deep Learning Networks and Its Application to Industrial Soft Sensors." *IEEE Transactions on Neural Networks and Learning Systems*. doi:10.1109/TNNLS.2022.3144162.

40. M. Zhang, S. Kim, P. Y. Lu, and M. S. Soljačić (2022) "Deep Learning and Symbolic Regression for Discovering Parametric Equations." *arXiv*, July. doi:10.48550/arxiv.2207.00529.

41. A. Thakkar and R. Lohiya (2021) "Analyzing Fusion of Regularization Techniques in the Deep Learning-Based Intrusion Detection System." *International Journal of Intelligent Systems*, 36(12): 7340–7388. doi:10.1002/INT.22590.

42. X. Bai et al. (2021) "Explainable Deep Learning for Efficient and Robust Pattern Recognition: A Survey of Recent Developments." *Pattern Recognition*, 120: 108102. doi:10.1016/J.PATCOG.2021.108102.

43. L. Deng et al. (2021) "Comprehensive SNN Compression Using ADMM Optimization and Activity Regularization." *IEEE Transactions on Neural Networks and Learning Systems*, doi:10.1109/TNNLS.2021.3109064.

44. S. Narduzzi, S. A. Bigdeli, S.-C. Liu, and L. A. Dunbar (2022) "Optimizing the Consumption of Spiking Neural Networks with Activity Regularization." Paper presented at IEEE Conference on Acoustics, April, pp. 61–65. doi:10.1109/ICASSP43922.2022.9746375.

45. M. Decuyper, M. Stockhoff, S. Vandenberghe, and X. Ying (2019) "An Overview of Overfitting and its Solutions." *Journal of Physics: Conference Ser*ies, 1168(2): 022022. doi:10.1088/1742-6596/1168/2/022022.

46. C. Zhang, O. Vinyals, R. Munos, and S. Bengio (2018) "A Study on Overfitting in Deep Reinforcement Learning." *arXiv*, April. doi:10.48550/arxiv.1804.06893.

47. H. K. Allamy and R. Z. Khan (2014) "Methods to Avoid Over-Fitting and Under-Fitting in Supervised Machine Learning (Comparative Study)." *Computer Science, Communication & Instrumentation Devices*, December: 163–171.

48. N. Srivastava, G. Hinton, A. Krizhevsky, and R. Salakhutdinov (2014) "Dropout: A Simple Way to Prevent Neural Networks from Overfitting." *Journal of Machine Learning Research*, 15: 1929–1958.

49. S. Salman and X. Liu (2019) "Overfitting Mechanism and Avoidance in Deep Neural Networks." *arXiv*, January.,doi:10.48550/arxiv.1901.06566.

50. J. Shunk (2022) "Neuron-Specific Dropout: A Deterministic Regularization Technique to Prevent Neural Networks from Overfitting & Reduce Dependence on Large Training Samples." *arXiv*, January.,doi:10.48550/arxiv.2201.06938.

51. F. Yang, H. Zhang, and S. Tao (2022) "Simplified Multilayer Graph Convolutional Networks with Dropout." *Applied Intelligence*, 52(5): 4776–4791. doi:10.1007/S10489-021-02617-7/TABLES/14.

52. A. Rusiecki (2022) "Batch Normalization and Dropout Regularization in Training Deep Neural Networks with Label Noise." *Lecture Notes on Networks Systems*, 418 LNNS: 57–66. doi:10.1007/978-3-030-96308-8_6/COVER.

53. S. Kumawat, G. Kanojia, and S. Raman (2022) "ShuffleBlock: Shuffle to Regularize Convolutional Neural Networks." Paper presented at 2022 National Conference on Communications, May, pp. 36–41. doi:10.1109/NCC55593.2022.9806750.

54. M. Bentoumi, M. Daoud, M. Benaouali, and A. Taleb Ahmed (2022,) "Improvement of Emotion Recognition from Facial Images Using Deep Learning and Early Stopping Cross-Validation." *Multimedia Tools and Application*, April: 1–31. doi:10.1007/S11042-022-12058-0/TABLES/13.

55. R. Shen, L. Gao, and Y.-A. Ma (2022) "On Optimal Early Stopping: Over-informative versus Under-informative Parametrization." *arXiv*, February.,doi:10.48550/arxiv.2202.09885.

Chapter 5

Motion Images Object Detection Over Voice Using Deep Learning Algorithms

P. L. Chithra and J. Yasmin Banu

5.1 Introduction

5.1.1 Self-Driving Vehicles

Self-driving vehicles are cars or trucks without human drivers to securely operate the vehicle. Conjointly referred to as autonomous or driverless cars, they combine sensors and computer code to regulate, navigate, and drive the vehicle.

5.1.2 Levels of Autonomy

Different cars are capable of various levels of self-driving, and area units are usually represented by researchers on a scale of 0–5 (Figure 5.1):

> Level 0: All main systems units are controlled by humans.
> Level 1: Certain systems, like control or automatic braking, are also controlled by the vehicle, one at a time.

DOI: 10.1201/9781003424550-5

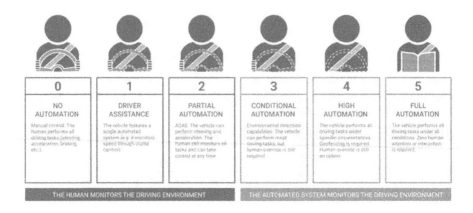

Figure 5.1   Levels of driving automation.

> Level 2: The car offers a minimum of two concurrent machine-driven functions, such as the acceleration and steering, however, it needs humans for safe operation.
>
> Level 3: The car will manage all safety-critical functions in limited conditions, however, the human driver can take control of the vehicle if alerted.
>
> Level 4: The car is fully autonomous in some driving circumstances, though not all.
>
> Level 5: The car is totally capable of self-driving in every situation.

5.1.3 Digital Image Processing

Digital image processing, or image processing, is a subdivision of computer vision. It deals with improving or highlighting and understanding pictures through various algorithms. An image might even be printed as a two-dimensional performance or two-dimensional array, where x and y squares measure spatial (plane) coordinates, and additionally the amplitude of any combination of coordinates (x, y) is known as the intensity or grey level of the image at that time. When x, y, and additionally the intensity values of squares measure all finite, distinct quantities, therefore, the image is referred to as a digital image. Note that a digital image consists of a finite sort of element, and each of the elements has a specific price at a definite location. These elements of square measures are named image parts, and pixels. This component of the pixel part is usually denoted as the element of a digital image.

5.1.3.1 Image Pre-Processing

The pre-processing technique is used to remove or suppress unwanted noise and improve the image quality in such a way by using image processing techniques called smoothing and sharpening. The other image pre-processing techniques used are rotation, scaling, zooming, resizing, converting images to grayscale images, and transforming.

5.1.4 Deep Learning

Deep learning (also called deep structured learning) may be a part of a broader family of machine learning ways that support artificial neural networks with illustration learning. Learning may be supervised, unattended, and reinforced. Deep learning is also an assortment of algorithms galvanized by the workings of the human brain as a method to process information and create patterns to be used in the higher cognitive process. Deep learning is based on a model design known as Artificial Neural Network. Deep learning has achieved enormous success in computational science, and its algorithms are broadly used in industries that solve advanced problems. Each of the deep learning algorithms uses numerous varieties of neural networks to accomplish determined tasks. Deep learning algorithms train computers by learning from examples. Industries such as healthcare, e-commerce, entertainment, and advertising commonly use deep learning.

5.1.5 Convolution Neural Networks

A convolution neural network (CNN) is one of the deep learning algorithmic programs. Neural networks square measure was impressed by how the human brain works, consisting of the many layers of interconnected neurons that act together. After all, if our brain can learn to comprehend, why can't a synthetic brain do the same? Neural networks enable computers to acknowledge the advanced patterns necessary for the particular task it needs to do. During this case, the aim is to show a machine how to ascertain and perceive the setting that it is in. The most effective approach to teach a vehicle to ascertain is to use a special style of neural network known as a convolutional neural network. CNN is employed in computer vision as a result of its superb ability to acquire spatial knowledge. This suggests that if we have an image of an individual, even though we tend to rotate it, move it around, squeeze and stretch it, the CNN would still acknowledge that it's an individual. The key behind the ability of CNNs is that they use special layers known as convolutional layers that extract patterns from the image. The initial convolutional layers acknowledge low-level patterns like edges and corners. Because the patterns progress through additional convolutional layers, the detected patterns become much higher in complexity. For instance, the convolutional layers used to observe individuals might go from edges to shapes to limbs to individuals as a full person. Convolutional neural networks by themselves use square measure primarily employed in image classification when given a picture, the network can accurately assign a given category.

5.1.6 Computer Vision

Computer vision means the extraction of data from pictures, text, videos, etc. Generally, computer vision strives to imitate human vision. It is a subdivision of Artificial Intelligence that collects data from digital pictures or videos and splits them to describe the attributes. The complete method involves image obtaining, viewing, reviewing, distinguishing, and extracting data. This in-depth process helps computers to know any visual content and consequently act upon it. Computer vision projects translate digital visual content into precise descriptions to assemble multidimensional knowledge. This knowledge is then turned into a computer-readable language to assist the decision-making method. The main objective of this wing of AI is to train machines to gather information from pictures.

5.1.6.1 Applications of Computer Vision

Computer vision is used in the following tasks:

- *object detection*: the position or place of the object.
- *object recognition*: the objects within the image, and their positions.
- *object classification*: the massive model that the object lies in.
- *object segmentation*: the pixel's integration to that object.

The field of AI has seen dramatic changes over the past few years resulting in several new techniques. Computer vision is one such field that has gained momentum in recent times that aims at instructing machines to study and interpret the visual world. Computer vision deals with numerous difficult cases ranging from classifying pictures to recognizing faces. One such challenge that might be addressed these days is object detection.

5.1.7 Object Detection

Object detection is a technique that incorporates two tasks: object classification and object localization. It is a model trained to observe the presence and place of multiple categories of objects. This could be used on static pictures or maybe in real time on videos. Object detection locates the thing and classifies it into completely different categories and localizes it by drawing boundary boxes around it. There are several use cases for object detection. As an example, self-driving or automated driving must be ready to determine different objects on the road while driving. To handle these use cases, various progressive algorithms are getting used to observing objects, e.g., R-CNN, Fast R-CNN, quicker R-CNN, Mask R-CNN, SSD, YOLO, etc. YOLO

(YOU ONLY LOOK ONCE) is considered one of the foremost powerful object detection algorithms, designed by Joseph Redmon and three others.

5.1.8 YOLO: You Only Look Once

Many object detection models grasp and process the image multiple times to be able to detect all the objects existing inside the pictures. But YOLO, as the name suggests merely looks at the item once. It applies one pass to the full image and predicts the bounding boxes and their category chances or possibility. This makes YOLO a superfast real-time object detection algorithm rule. One convolutional neural network at the same time predicts multiple bounding boxes and class probabilities for those boxes. YOLO uses features from the complete image to predict each boundary of the box, in real time, and its classes, which it does simultaneously. Similar to humans, YOLO will almost immediately recognize where and what objects are inside a given image. Once running on an image, YOLO first divides the image into associate SXS inside each grid cell, then YOLO can predict the locations, sizes, and associated confidence scores of the planned range of bounding boxes, essentially predicting the category and potential place wherever an object is. If the centre of the associate object falls in the grid cell, then the bounding

Figure 5.2 Input image divided into grid cells.

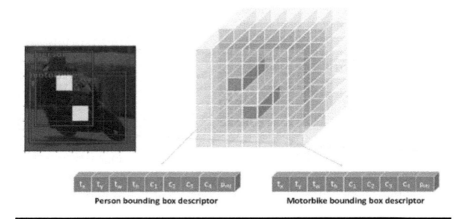

Figure 5.3   Predicted feature map.

boxes of that grid cell area unit accurately locate and predict that object (Figures 5.2 and 5.3).

Each bounding box can have five predictions: x coordinate, y coordinate, width, height, and confidence. The calculated confidence score indicates how certain the model thinks there's a category inside the bounding box, and the correct way it thinks that category fits within the box uses a metric known as intersection over the union. Intersection over the union is employed in object detection as a result of comparing the bottom true bounding box with the anticipated bounding box. By dividing the region of overlap with the region of the union, we've got a concept that rewards serious overlapping and penalizes inaccurate bounding box predictions. The bounding box's goal is to ultimately confine the item inside the image as accurately as possible and intersection over union may be a nice metric to do this. When a picture is run through YOLO, it outputs predictions to the associated S X S X (B* 5+C) tensor where every grid cell predicts the situation and confidence legion B quantity of bounding boxes across C quantity of categories. Ultimately, we tend to find ourselves with tons of bounding boxes – most of those are digressive. To filter the proper boxes, the bounding boxes with an expected category that meets a particular confidence score can then be unbroken. This enables all relevant objects inside the image to be isolated. Figure 5.4 gives an overview of the proposed method.

5.2 Literature Review

T.H. Jung et al. [1] designed a system that detects only persons in the agricultural environment. A four-channel frame camera was mounted on the four sides

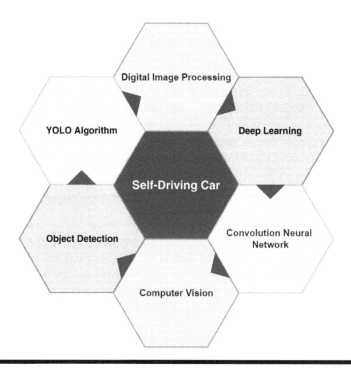

Figure 5.4   Overview of the proposed method.

(left, right, front and rear) of the tractor, and the data was collected from the entire surrounding. The Nvidia Jetson Xavier platform is used as an experimental system capable of performing high-speed GPU-based parallel processing to avoid resource problems. CNN [2] has several object detection algorithms, like R-CNN [3], Fast R-CNN [3], Faster R-CNN [3], Mask R-CNN, Over Feat, Single Shot Detector (SSD), and YOLO series. YOLO is considered the best network in this chapter, because it is capable of obtaining high-accuracy detection and can run in real time. YOLO (You Only Look Once), as the name implies, needs only one forward propagation through a neural network for object detection. YOLO [3] uses CNN for object detection. This chapter adopted the YOLOv3 [4] network for person recognition and some changes were made to apply the proposed system.

In a study [5] a pedestrian object detection system was set up using YOLOv3. The YOLO algorithmic rule is trained on visualization in only one category: the person category. It has achieved great precision and recall in comparison to plain YOLOv3. In a Lidar-based LS-R-YOLOv4 approach, a paper used a device purpose cloud map to measure data with deep learning to discover and section objects [6]. The primary information set was KITTI which provides depth data information for measuring a device segmentation (Lidar Segmentation) of objects obtained through measuring device purpose clouds and then done on pictures. Later in the YOLOv4 [7], a neural network region of interest (ROI) detects the objects as final object detection.

Tiny target detection supported the improved YOLO network approach [8]. The paper chiefly introduces the small object detection of road information supported by the proposed YOLO network model optimized by YOLOv3 network [4], which uses a twin feature extraction network structure. The backbone network is given a special feature extraction auxiliary network of the receptive fields. The feature information fusion of the auxiliary network and additionally the backbone network adopted the attention mechanism. It focuses on the effective feature channels, suppresses the invalid information channels, and improves the network method efficiency.

Object detection, using a modified YOLOv1 [9] neural network, is a modified network proposed for object detection. The old loss function of YOLOv1 [7] is modified. The loss function of the original YOLOv1 [10] takes the same error for large and small objects. So the prediction of neighbouring objects is unsatisfactory. Detecting two objects in the same grid becomes a problem, so the new loss function is more flexible and optimizable, compared to the old loss function and in a network error.

P.L. Chitra and Gomathi [11] proposed a method that detects a pedestrian with the lane detection method which is essential for self-driving vehicles using the YOLOv3 algorithm.

Handalage 12] reviewed the fundamental structure of the CNN algorithms, and an overview of YOLO's real-time object detection algorithm. The CNN design models will take away highlights and see objects in every given image. Here the authors used the YOLOv2 model to convey the summary of this work.

Thakkar et al. [13] reported on a custom image dataset being used and trained for six classes using YOLO to track and sort. Recognizing a vehicle or a pedestrian in an ongoing video is helpful for traffic analysis.

Pal and Chawan [14] gave a brief introduction about deep learning and an object detection framework, in which they clearly explained CNN and its latest algorithms for object detection.

After going through the deep network Lan et al. [15] improved the network structure of the YOLO algorithm and proposed a new network structure YOLO-R. There was some loss of pedestrian information, which caused the disappearance of gradients, causing inaccurate pedestrian detection.

5.3 Proposed Methodology

Road accidents are quite common today. Particularly in large cities, there are numerous modes of transport. Moreover, the roads have become narrower and

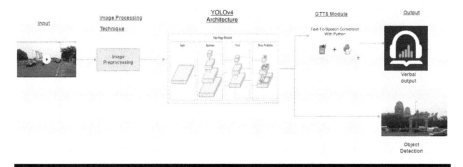

Figure 5.5   Proposed architecture.

also the cities have become more crowded. Object detection is a significant part of driverless vehicles. This work is projected to find completely different categories which are trained using the YOLOv4 design besides the conversion of object labels (text) into speech or audio responses for the visually impaired person to grasp unfamiliar surroundings. This works by exploiting the YOLOv4 algorithmic program that runs through a variation of a very advanced convolutional neural specification referred to as the Dark net with OpenCV and Google Text to Speech. The methodologies for multiple object detection with audio output are shown in Figure 5.5.

5.3.1 Proposed Architecture

In this proposed architecture, the input data are collected using the video which was taken during traffic and then converted into frames. This system focuses on recognizing and identifying the specified five classes (person, car, bus, bike, auto) which appear in the traffic environment. The five classes are considered to be the objects in our datasets. Further, these frames are processed using the following workflow as shown in Figure 5.6 and detailed here.

The flowchart in Figure 5.6 of the proposed method starts by receiving input images and processing them on the YOLO network to detect objects. Once it has done the detection, it converts the object label into speech to produce audio output until it terminates or stops the execution.

5.3.2 Labelled Input Image

The traffic signal video is converted into image frames which are later labelled using the labelling tool.

Step 1: The frames are input images, the labelling tool has been used to annotate those images.

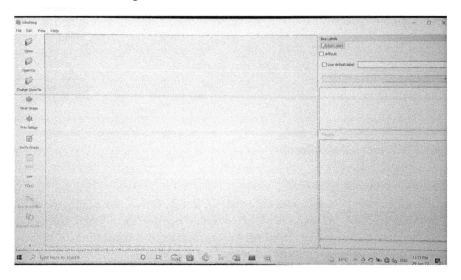

Step 2: Creation of bounding box using the labelling tool.

Step 3: Save an annotation file with the coordinates.

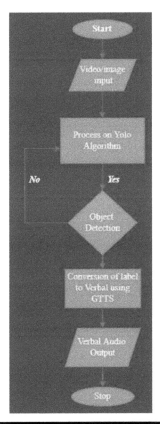

Figure 5.6   Flowchart of the proposed method.

5.3.3 Stages of the Proposed Method

Step 1: Video acquisition: This is a process of capturing motion video using a high-resolution camera. In this work, the 30 seconds of traffic video have been captured in the format of mp4 along with different angles and different directions during the traffic time.

Step 2: Image frames: This is a process of converting videos into image frames from motion video. Later these image frames are used for training and testing the algorithm.

Step 3: YOLOv4 object detector: This is the process to detect an object from the given image frames and predict the object correctly.

Step 4: Object bounding boxes: In order to recognize the various objects such as persons, cars, buses, bikes, and autos, bounding boxes with labelled names are used so that they can easily be identified in the detection of an object.

Step 5: Detected object: Finally, with the predicted bounding boxes are detected objects of various images.

Step 6: Audio segment: This converts the predicted object label (text) into speech or audio responses.

5.3.4 YOLOv4 Architecture

Already we have discussed YOLO, which is one of the most famous object detection algorithms due to its speed and accuracy. YOLOv4 is intended for this process, based on recent analysis findings, using CSPDarknet53 as a backbone, SPP (spatial pyramid pooling) and PAN (path aggregation network), which is referred to as "the Neck", and YOLOv3 for "the Head". The network contains 53 convolution layers with the sizes of 1x1 and 3x3, and every convolution layer is connected with a batch normalization (BN) layer and a Mish activation layer. SPP may be a methodology to exploit both fine and coarse information by simultaneously pooling on multiple kernel sizes (1, 5, 9, 13). PAN is a technique that leverages information in layers close to the input by conveying features from completely different backbone levels to the detector. This model contains 53 convolution layers, each followed by a batch normalization layer and Leaky ReLU activation. Predicting the output contains the centre of the ground truth box of an object, which predicts the image. The centre of an image is responsible for the detection of objects, and also predicting through probabilities that the box contains a certain class.

5.3.4.1 Intersection over Union (IoU)

Intersection over Union (IoU) is employed once mean average precision (mAP) is found. This is a number from zero to one that specifies the quantity of overlap between the predicted and ground truth bounding box.

- an IoU = 0, then no overlap between the boxes.
- an IoU = 1, then completely overlapping that is the union of the boxes is the same as their overlap.

5.3.4.2 mAP

Mean average precision (mAP) is employed by collecting a set of predicted object detections and a set of ground truth object annotations.

- For every prediction, IoU is computed with reference to each ground truth box enclosed in the image.
- These IoUs are then thresholded to some value (generally between 0.5 and 0.95) and predictions are matched with ground truth boxes with respect to a greedy strategy (i.e. highest IoUs are matched first).
- A precision recall (PR) curve is then generated for every object class and therefore the average precision (AP) is computed. A PR curve takes into consideration the performance of a model with reference to true positives, false positives, and false negatives over a spread of confidence values.
- The mean of the AP of all categories classes is the mAP.

5.3.4.3 Precision

Precision is the ratio of the number of true positives to the total number of positive predictions. For example, if the model detected 100 cars, and 90 were correct, the precision is 90 per cent.

5.3.4.4 Recall

Recall is the ratio of the number of true positives to the total number of actual (relevant) objects. For example, if the model correctly detects 75 cars in an image, and there are actually 100 cars in the image, the recall is 75 per cent.

The F1 score is a weighted average of the precision and recall. Values range from 0 to 1, where 1 means the highest accuracy.

5.3.4.5 Non-Maximum Suppression (NMS)

Non-maximum suppression (NMS) may be the final step that is used in various computer vision tasks. It is a category of algorithms to choose one entity (e.g., bounding boxes) out of many overlapping entities. The purpose of NMS is to decide the best bounding box for an object and reject or "suppress" all other alternative bounding boxes. The NMS takes two things into consideration:

1. the objectiveness score of the model;
2. the overlap of the bounding boxes.

5.3.5 Google Text-to-Speech

Google Text-to-Speech, a Python library, and a user interface tool are used to interface with Google Translates text-to-speech application programming interface (API). It features versatile pre-processing and tokenizing. Text-to-Speech generates raw audio information of natural, human speech. That is, it creates audio that appears like someone talking. When you send a synthesis request to Text-to-Speech, you need to specify a voice that 'speaks' the words. Text-to-Speech features a wide variety of customized voices to use. The voices vary in language, gender, and accent (for some languages). For instance, we are able to produce audio that mimics the sound of a female English speaker with a British accent. We are able to additionally convert an equivalent text into a distinct voice, say, a male English speaker with an Australian accent.

5.3.6 YOLOv5 Models' Approach to Checking Performance

Object detection is a computer vision technique for locating instances of objects in images or videos, along with their classes and bounding boxes.

5.3.6.1 Design of the Proposed Method

This method uses the YOLOv5 version along with its four kinds of models to check the performance (Figure 5.7). They are YOLOv5s, Yolov5m, Yolov5l, and Yolov5x. Abbreviated as small, medium, large, and extra-large.

- *Capturing the decoding video file*: we have captured the video using Video Capture object and then capturing has been initialized so every video frame is decoded (i.e., converted into a sequence of images).
- *Pre-processing*: The converted image frames have been preprocessed into a particular size. In this method, we have used the image size of 416x416 with the channel as 3 for colour images.
- *Labelling*: The pre-processed images are labelled using the labelling tool to annotate the images. As the YOLO object detection model is a supervised learning algorithm, it requires labelled datasets. YOLO versions always use a text file for their annotation format with (.txt) as an extension.
- *YOLOv5 object detection*: In this version, we have taken YOLOv5s, YOLOv5m, YOLOv5l, and YOLOv5x models for the detection of an object.
- *Object detection*: Predicted bounding boxes with confidence scores are detected objects of various images.

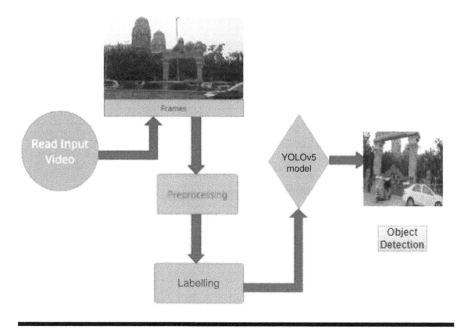

Figure 5.7   Flowchart of the method.

Here we have implemented YOLOv5 based on the PyTorch framework which uses a .yaml file for configuration. One of the most popular Python-based deep-learning libraries is PyTorch and Tensorflow. YOLOv5 s, m, l, and x models output structure displayed in Tensorboard in Section 5.4. PyTorch is one of the foremost recent deep learning frameworks, developed by Facebook. It is gaining popularity because of its simplicity of use, and dynamic computational graph.

5.4 Experimental Results

To execute the proposed techniques, an experimental set-up was proposed, to detect people, bus, car, auto, and bike. This work takes 200 frames in various video images to prove experimental results. For object detection, different sets of images are taken freely in different scenes, with different lighting conditions, and create defects in the bounding box using the YOLOv4 algorithm.

The outcome of the problem is the calculated average precision (AP) and the mean average precision (maP) in overall detections and achieves better results (Table 5.1).

5.4.1 Detection of mAP Performance

Figure 5.8 shows the mAP and loss chart.

Table 5.1   Detection Evaluation Metrics

Metrics	Values
Precision (%)	71
Recall (%)	67
Conf_thresh	0.25
mAP (%)	71.5
IoU	50.86
F1-score (%)	69
Average loss	0.7

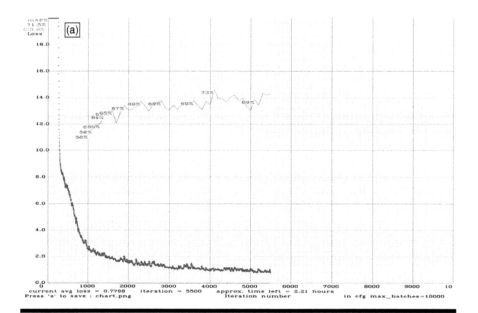

Figure 5.8   m(a) mAP and loss chart, (b) YOLOv4 Performance Outcomes with Audio.

Figure 5.8 (continued)

Figure 5.8 (continued)

Figure 5.8   (continued)

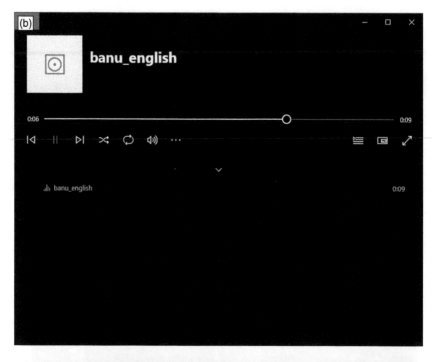

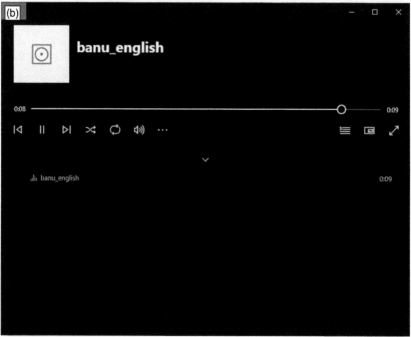

Figure 5.8 (continued)

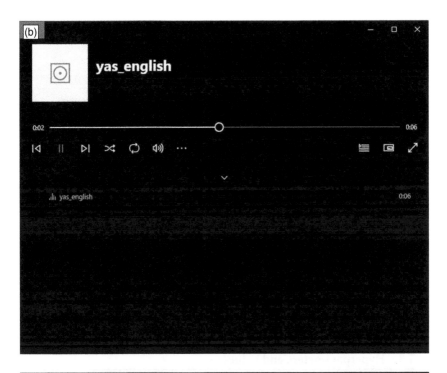

Figure 5.8 (continued)

Table 5.2 Comparison of YOLOv5 Models (s, m, l, x) (%)

Models	Precision (%)	Recall (%)	$_{mAP}test$ (%)	F1-measure (%)	Speed of inference (s)
YOLOv5s	90.4	74.5	78.5	81.68	0.023s
YOLOv5m	89.7	78.3	83.0	83.61	0.046s
YOLOv5l	91	76	80	82.82	0.071s
YOLOv5x	92	78.4	83.4	84.42	0.039s

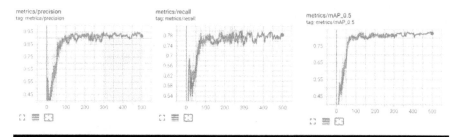

Figure 5.9 Comparison of performance metrics, during training for YOLOv5 l (light) and YOLOv5 x (dark).

Table 5.3 Hyperparameters and Values

Hyperparameter	Value
Activation	Leaky ReLU, Sigmoid
Decay	0.0005
Iteration	500
Learning rate (Initial)	0.01
Learning rate (final)	0.1
Momentum	0.937
Image size	416
Channel	3
Number of classes	5

5.4.2 Experimental Outcome Results

The experimental results are shown in Table 5.2 and Figure 5.9.

We have compared the YOLOv5 version with their different models like s, m, l, x. Table 5.3 shows the hyperparameter values set during the training of the network for 500 iterations.

5.4.3 Predictions

Figures 5.10 a–d show the detection of YOLOv5 based on different models. The model YOLOv5 x has achieved 98 per cent of predictions accurately whereas we can see the model has achieved 93 per cent of the lowest predictions.

1. YOLOv5 x
2. YOLOv5 l
3. YOLOv5 m
4. YOLOv5 s

Figure 5.10 **Testing detected results based on YOLOv5 algorithm: (a) YOLOv5 x, (b) YOLOv5 l, (c) YOLOv5 m, (d) YOLO v5 s.**

Figure 5.10 (continued)

Figure 5.10 (continued)

Figure 5.10 (continued)

5.5 Conclusion

In this study, the five classes were taken as an object to detect if it appears in the image frames that are mainly used for the self-driving vehicles in the traffic environment. This work represents object detection used with the YOLO architecture. YOLOv4 is the latest and fastest algorithm with a highly accurate rate of true positive predictions. The width and height of predicted bounding boxes and IOU methods were used to identify the real-time detection. In future, the algorithm can be fine-tuned and increases the number of classes in different weather conditions. In our future work, the network architecture along with the weight file will be created rather than using the pre-trained model with different numbers of objects and increasing the number of classes in our datasets.

References

1. Taek-Hoon Jung, Benjamin Cates, In-Kyo Choi, Sang-Heon Lee, and Jong-Min Choi (2020) "Multi-Camera-Based Person Recognition System for Autonomous Tractors". http://dx.doi.org/10.3390/designs4040054
2. https://towardsdatascience.com/a-comprehensive-guide-to-convolutional-neural-netwo rks- the-eli5-way-3bd2b1164a53
3. https://towardsdatascience.com/r-cnn-fast-r-cnn-faster-r-cnn-yolo-object-detect ion- algorithms-36d53571365e
4. Joseph Redmon and Ali Farhadi (2018) "Yolov3: An Incremental Improvement". *arXiv* 1804.02767v1
5. https://towardsdatascience.com/review-yolov2-yolo9000-you-only-look-once-obj ect- detection-7883d2b02a65
6. Yu-Cheng Fan, Yelamandala Chitra, Ting-Wei Chen, and Chun-Ju Huang (2021) "Real-Time Object Detection for LiDAR Based on LS-R-YOLOv4 Neural Network". *Journal of Sensors*, 1–11. doi:10.1155/2021/5576262.
7. A. Bochkovskiy, C. Y. Wang, and H. Y. M. Liao (2020) "Yolov4: Optimal Speed and Accuracy of Object Detection". *arXiv*. https:// arxiv.org/abs/2004.10934.
8. Qiwei Xu, Runzi Lin, HanYue, Hong Huang, Yun Yang, and Zhigang Yao (2020) "Research on Small Target Detection in Driving Scenarios Based on Improved Yolo Network". *IEEE Access*. doi:10.1109/ACCESS.2020.2966328.
9. Tanvir Ahmad, Yinglong Ma, Muhammad Yahya, Belal Ahmad, Shah Nazir, Amin Haq, and Rahman Ali (2020) "Object Detection through Modified YOLO Neural Network". *Scientific Programming*. doi:10.1155/2020/8403262.
10. J. Redmon, S. Divvala, R. Girshick, and A. Farhadi (2016) "You Only Look Once: Unified, Real-Time Object Detection". Paper presented at 2016 IEEE Conference on Computer Vision and Pattern Recognition (CVPR), Las Vegas, NV, USA, pp. 779–788.
11. P. L. Chithra and V. Gomathi (2021) "Video Analysis Based on Pedestrian Object Detection using YOLOv3". Available at: www.researchgate.net/publication/363211 040
12. Upulie Handalage and Kuganandamurthy Lakshini (2021) "Real-Time Object Detection Using YOLO: A Review". Preprint. doi:10.13140/RG.2.2.24367.66723.
13. Heet Thakkar, Noopur Tambe, Sanjana Thamke, and Vaishali Gaidhane (2020) "Object Tracking by Detection Using YOLO and SORT". *International Journal of Scientific Research in Computer Science, Engineering and Information Technology*, 30 March: 224–229. doi:10.32628/CSEIT206256.

14. Shubham Pal and Pramila Chawan (2019) "Real-Time Object Detection Using Deep Learning: A Survey". *International Research Journal of Engineering and Technology*, 6: 2395–056.

15. Wenbo Lan, Jianwu Dang, Yangping Wang and Song Wang (2018) "Pedestrian Detection Based on YOLO Network Model". Paper presented at IEEE International Conference on Mechatronics and Automation. 978-1-5386-60751/18/$31.00 IEEE.

Chapter 6

Diabetic Retinopathy Detection Using Various Machine Learning Algorithms

P. K. Nizar Banu and Yadukrishna Sreekumar

6.1 Introduction

Diabetes is said to be a metabolic disease which causes high blood sugar. India is said to be the diabetic capital of the world by 2030 with over 80 million people affected. So it is necessary to diagnose and treat diabetes in the early stages. Diabetic retinopathy is another condition which is connected to diabetes. Retinopathy is a condition which develops in the eye, which if not treated at early stages, could lead to permanent blindness.

Diabetic retinopathy, also known as diabetic eye disease, is a medical condition in which damage occurs to the retina due to diabetes mellitus. Diabetic retinopathy is the leading cause of blindness in the working-age population of the developed world (California Healthcare Foundation, 2015). According to the California Healthcare Foundation, diabetic retinopathy will affect over 93 million people. The US Centers for Disease Control and Prevention estimate that 29.1 million people in the US have diabetes and the World Health Organization estimates that 347 million people have the disease worldwide. If a person has long-standing diabetes, they are potentially a diabetic retinopathy sufferer. It is observed by the US Centers for Disease Control and Prevention, that around 40–45 per cent of Americans with diabetes have diabetic retinopathy, from early to advanced stages. It is difficult to provide effective treatment as the disease often shows few or no symptoms and it may lead to vision impairment. Detection of diabetic retinopathy is a manual and time-consuming process which requires a trained clinician to examine and evaluate digital colour fundus photographs of the retina. Normally, clinicians identify diabetic retinopathy by the presence of lesions associated with the vascular abnormalities caused by the disease. Though this approach sounds effective, it demands high resources.

Diabetic retinopathy happens when the small blood vessels get damaged and it damages the neurons of the retina. Retinal arteries are compressed with less retinal blood flow, wounding the neurons of the inner retina, the outer retina is associated with the changes in the visual function, harming the blood-retinal barrier, which may also lead to the leakage of blood constituents into the retinal neuropile, these are the initial changes that can be found in people affected with DR (https://en.wikipe dia.org/wiki/Diabetic_retinopathy). Effective diabetic retinopathy screening is required because diabetic retinopathy does not show any symptoms in the initial stages, and can cause blindness if it is not diagnosed and treated promptly (Antal and Hajdu, 2014; Narasimha-Iyer et al., 2006; Niemeijer et al., 2009; Osareh et al., 2009; Pires et al., 2013; Sil Kar and Maity, 2016; Taylor et al., 2009; Walter et al., 2002; Zhang et al., 2009).

Research on diabetic retinopathy detection has received good reviews using image classification, pattern recognition and machine learning. Automated methods of diabetic retinopathy detection are anticipated as the numbers of individuals with diabetes continue to grow steadily. Deep neural networks offer a great advantage of screening for DR from retinal images, in improved identification of DR lesions and risk factors for diseases, with high accuracy and reliability. In effect, convolutional neural networks (CNN) (a deep learning method) have been taught to recognize pathological lesions from images.

This research work discusses the automatic detection of diabetic retinopathy in retinal images using machine learning techniques. Also the work investigates the capability of images pre-retinopathy. In addition, a website has been constructed (Diabetes Web) which will be helpful in predicting diabetes as well as diabetic retin-opathy with the help of a blood sample details and images respectively. This research work uses colour fundus photography for analysis. Figure 6.1 shows the image of the fundus for diabetic retinopathy.

6.2 Background

6.2.1 Proliferative Diabetic Retinopathy

As the disease progresses, severe non-proliferative diabetic retinopathy enters an advanced or proliferative stage, where blood vessels proliferate/grow. The lack of oxygen in the retina causes fragile, new, blood vessels to grow along the retina and

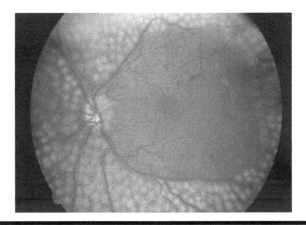

Figure 6.1   Fundus image – scatter laser surgery for diabetic retinopathy.

*Source: https://en.wikipedia.org/wiki/File: Fundus_photo_showing_scatter_
laser_surgery_for_diabetic_retinopathy_EDA09.JPG*

in the clear gel-like vitreous humour that fills the inside of the eye. Without timely treatment, these new blood vessels can bleed, cloud the vision, and destroy the retina (www.rxlist.com/diabetic_retinopathy/definition.htm). Fibrovascular prolif-eration can also cause tractional retinal detachment (Conrad Stöppler, 2021). The new blood vessels can also grow into the angle of the anterior chamber of the eye and cause neurovascular glaucoma. Non-proliferative diabetic retinopathy shows up as cotton wool spots, or microvascular abnormalities or as superficial retinal haemorrhages. Even so, the advanced proliferative diabetic retinopathy (PDR) can remain asymptomatic for a very long time, and so should be monitored closely with regular check-ups (www.rxlist.com/diabetic_retinopathy/definition.htm). A computer-aided screening system that uses machine learning classifiers such as k-nearest neighbour (kNN), support vector machine (SVM), Gaussian mixture model (GMM) and Adaboost in classifying retinopathy lesions from non-lesions are analysed in (Roychowdhury et al., 2014), Several methods were used for automatic retinopathy detection and diagnostic procedure of proliferative diabetic retinopathy (Randive et al., 2019). This study used fundus images with varying illumination and views to generate a severity grade for diabetic retinopathy. The work carried out by Alazzam et al. (2021) claims a radial basis machine shows good diagnostic accuracy and can be potentially used in screening for diabetic retinopathy. With the support of machine learning and advanced intelligence, the Diabetes Web would provide the essential support and guidance to predict both diabetes and diabetic retinopathy. The Diabetes Web, developed as an application in this research work,[1] is a website which will mainly benefit clinicians and ophthalmologists where they will be able to predict diabetes and diabetic retinopathy, such as identifying which stage of diabetes retinopathy the patient is in, based on the blood count and image of the retina respectively.

The main objectives of the Diabetes Web will be prediction of diabetes and diabetic retinopathy, which will be helpful for patients to start their treatment based on the results obtained through our website. Because of late diagnosis, diabetic retinopathy lead to permanent blindness and many other health issues. Hence this chapter addresses the following tasks:

- Identify machine learning models for diabetes prediction.
- Analyse different models for diabetic retinopathy detection.
- Present a website which will be helpful for predicting diabetes as well as levels of diabetic retinopathy.

The Diabetes Web is designed as an open-source web application where the users will be able to predict diabetes by giving the count of insulin, glucose content and age. Also, the users can upload images of the retina and get instant results for diabetic retinopathy (DR). Manual effort is reduced and the application is easily accessible from any device across platforms.

This research work is carried out as an analysis to serve the needs of ophthalmologists and lab technicians, and for rural areas (Tier 4 cities), with small dispensaries, who do not have medical professionals for an emergency but can afford a fundus camera. It also acts as a tool for doctors to prioritize cases based on the severity of the disease condition.

6.3 Applicability

Diabetes Web is a website where the users will be able to predict diabetes. If the user is found to be diabetic, then the users can check their diabetic retinopathy status as well on the same website. The users are also able to check on which level of DR they are, by uploading the fundus image of the retina. This website will be very useful for clinicians as well as ophthalmologists. They can check the tests which they are doing manually and confirm both diabetes as well as diabetic retinopathy. Diabetes Web can be considered one of the most important tools for doctors to prioritize their patients based on their criticality. Also, Diabetes Web will be more relevant in Tier 2 areas where patients are unable to afford ophthalmologists or experts.

The main aim of Diabetes Web is to predict whether the patient is diabetic or not. Then there is another feature which could help ophthalmologists to detect blindness before it happens. Millions of people suffer from diabetic retinopathy, the leading cause of blindness among working aged adults ((APTOS, 2019). In India, Diabetes Web will help to detect and prevent this disease among people living in rural areas where medical screening is difficult to conduct. APTOS (Asia Pacific

Tele-Ophthalmology Society Symposium) has provided eye fundus images under a variety of imaging conditions with labels to detect the DR level in retinal images. Each image is rated on a scale of 0–4: 0 = No DR; 1 =Mild; 2 = Moderate; 3 = Severe; 4 = Proliferative DR. You can detect the locations of DR on retinal images and scale DR levels from 0–4. The website has certain non-functional requirements, such as the website should be highly responsive and fast. It should be user-friendly and simple. There should be only a single retina per image. Multiple retinas on a single page are not allowed. The image which is going to be uploaded should be a jpeg or png or bmp file and the maximum size of the image should be less than 50 mb. Basically, there are different kinds of users who may access the system:

- *Clinicians*: Clinicians can make use of Diabetes Web to check the diabetes results of the patients. They can reconfirm their results in comparison with the results they obtained through Diabetes Web.
- *Ophthalmologists*: Ophthalmologists can make use of Diabetes Web to predict the different levels of DR by uploading the fundus image of the patient's retina.
- *Doctors*: Doctors can also make use of this website for prioritizing their patients based on the severity of their DR.

Figure 6.2 shows the flow of the research work carried out. Figure 6.2 helps to explain the sub-systems making up the system and the framework for sub-system control and communication. From Figure 6.2, we can see that the user will be giving the input (in the form of an image which will contain the images of the retina) through our user interface.

With the help of Flask microservices we put the image into our python backend. From there the actual computation starts. As the first stage of processing we take the image into the image processing unit. In that module we extract the necessary information from the image, then, using a model, we order the extracted symbols

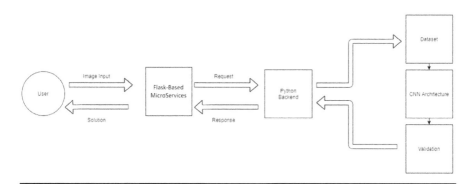

Figure 6.2   Framework of the research work.

and make that a digital equation. Then we have to confirm that with the user, so that it is the same equation which is uploaded by the user. After receiving confirmation from the user, we can put that digital equation into our pretrained model and obtain the results. That result will be moved to the python backend. Again, with the help of Flask microservices we can give that response to the user through our user interface.

The visual part of our web-based computer application is the part through which a user interacts with Diabetes Web. It determines how commands are given to the system or the program and how data is displayed on the screen. Our website will be a single page website where the user can input the images onto our site. Figures 6.3 and 6.4 show two pages of the dashboard of the Diabetes Web.

After uploading the image as an input to the system, the user will be able to see the result on the same page. Also, the user will be able to download the solution as a pdf, by clicking the print button. Figure 6.5 shows the diabetes prediction page. On this page, the user can input the necessary basic details to the site and get an instant result (diabetic or not) on the same page. Figures 6.6 and 6.7 show the diabetes retinopathy prediction page and how to choose an image.

6.4 Diabetic Retinopathy Detection

6.4.1 Vertical and Horizontal Flip in Image Augmentation

An image flip reverses the rows or columns of pixels in the case of a vertical or horizontal flip respectively. The flip augmentation is specified by a Boolean horizontal flip or vertical flip argument to the Image Data Generator class constructor. For photographs like the bird photograph, horizontal flips may make sense, but vertical flips would not. For other types of images, such as aerial photographs, cosmology photographs, and microscopic photographs, vertical flips make sense (Gondhalekar, 2020).

6.4.2 Random Rotation Augmentation

A rotation augmentation randomly rotates the image clockwise by a given number of degrees from 0 to 360. The rotation will likely rotate pixels out of the image frame and leave areas of the frame with no pixel data that must be filled in. The example in https://machinelearningmastery.com/how-to-configure-image-data-augmentation-when-training-deep-learning-neural-networks/ demonstrates random rotations via the rotation range argument, with rotations to the image between 0 and 90 degrees. For random brightness augmentation, the brightness of the image can be augmented by either randomly darkening images, brightening images, or both. The aim is to allow a model to generalize across images trained on different lighting levels. This can be achieved by specifying the brightness range argument to the

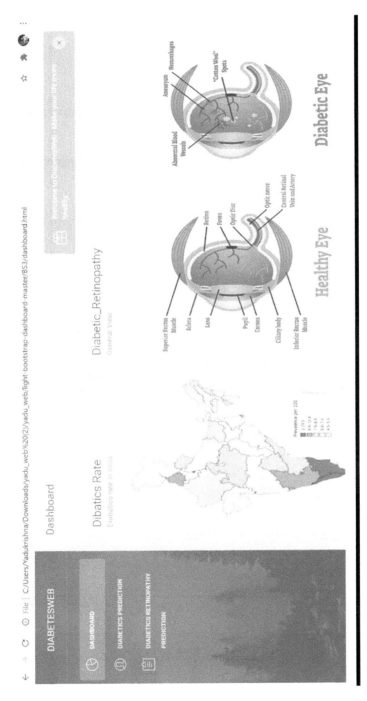

Figure 6.3   Dashboard of Diabetes Web.

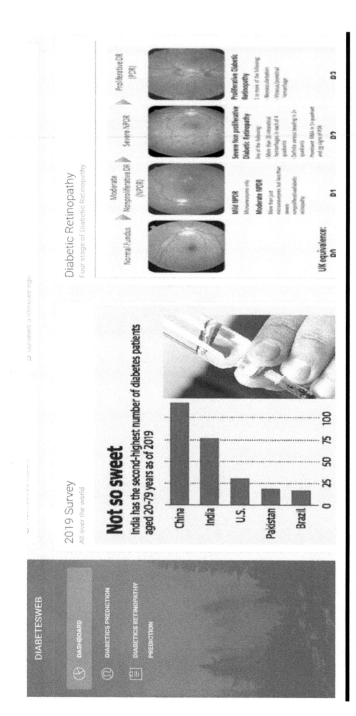

Figure 6.4   View of Diabetes Web dashboard page 2.

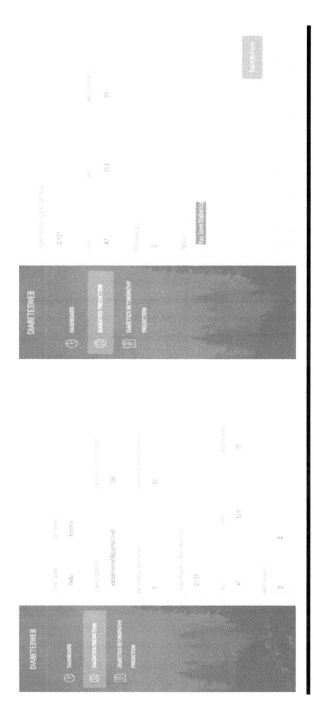

Figure 6.5   Diabetes prediction page.

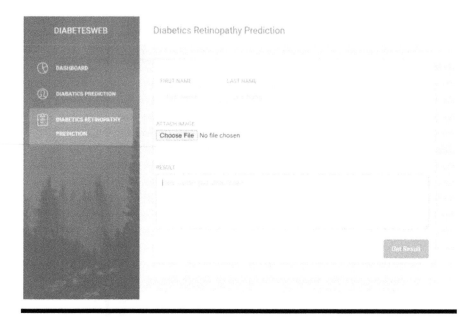

Figure 6.6   Diabetic retinopathy prediction page.

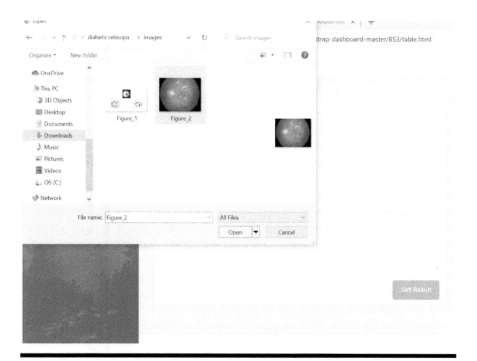

Figure 6.7   Diabetic retinopathy prediction page: choosing an image.

ImageDataGenerator constructor that specifies the minimum and maximum ranges as a float representing a percentage to select a brightening amount. Values less than 1.0 darken the image, e.g., 0.5, 1.0, whereas values greater than 1.0 brighten the image, e.g., 1.0, 1.5, where 1.0 has no effect on brightness. The example in https://machinelearningmastery.com/how-to-configure-image-data-augmentation-when-training-deep-learning-neural-networks/ demonstrates a brightness image augmentation, allowing the generator to randomly darken the image between 1.0 (no change) and 0.2 or 20 per cent.

DenseNet is composed of dense blocks. In those blocks, the layers are densely connected together. A dense block is a group of layers connected to all their previous layers (Douillard, 2018). A single layer consists of the following:

- Batch Normalization
- ReLU activation
- 3x3 Convolution

The authors found that the pre-activation mode (BN and ReLU before the Conv) was more efficient than the usual post-activation mode.

6.4.2.1 Transition Layer

Instead of summing the residual as in ResNet, DenseNet concatenates all the feature maps. It would be impracticable to concatenate feature maps of different sizes. Thus, in each dense block, the feature maps of each layer have the same size. However down-sampling is essential to CNN. A transition layer is made of Batch Normalization, 1x1 Convolution, Average pooling.

6.4.3 DenseNet for Semantic Segmentation

The typical segmentation architecture is composed of a down-sampling path responsible for extracting coarse semantic features and an up-sampling path trained to recover the input image resolution at the output of the model. Densely Connected Convolutional Networks (DenseNets) have shown excellent results on image classification tasks. The idea of DenseNets is based on the observation that if each layer is directly connected to every other layer in a feed-forward fashion, then the network will be more accurate and easier to train (Chablani, 2017). DenseNets are built from dense blocks and pooling operations, where each dense block is an iterative concatenation of the previous feature maps. This architecture can be seen as an extension of ResNets, which performs an iterative summation of the previous feature maps. These small modifications have interesting implications, such as parameter efficiency, implicit deep supervision and feature use. The characteristics of DenseNets make them a very good fit for semantic segmentation as they naturally skip connections and multi-scale supervision, in the following manner:

- visualizing first 10 test images (Figure 6.8)
- visualizing testing images by training of DenseNet121 and predicting the scale of DR for each label (Figure 6.9)
- DenseNet121 Architecture for transfer learning (Figure 6.10) (Jee et al., 2021).

We analysed around six models for diabetes prediction and also analysed two models for DR detection.

6.4.4 Dataset Description

The system's dataset comes from the Mendeley Diabetes Types dataset. This information was gathered from Iraqi society at the Medical City Hospital's laboratory (the Specialized Center for Endocrinology and Diabetes-Al-Kindy Teaching Hospital). The data from the patients' file is extracted and entered into the database. Medical information and laboratory analysis make up the data. There are 103 patients with no diabetes, 53 with pre-diabetes, and 844 with diabetes in the database.

For diabetic prediction, the results are as shown in Table 6.1. Here the accuracy was high for the KNN (k nearest neighbours) algorithm, so we have connected KNN with the frontend.

For DR detection, the images are accessed from www.kaggle.com/competitions/diabetic-retinopathy-detection/data, and we have created a couple of models and the results are as shown in Table 6.2.

Here we obtained more accuracy for DenseNet so we have taken DenseNet model to connect the user interface.

6.5 Conclusion

In this chapter, a web application called Diabetes Web is introduced to analyse a patient's diabetes level. This website helps to identify the degree of diabetic retinopathy in a patient. Basic machine learning applications are used to detect if the patient is diabetic or not. Deep learning techniques such as CNN and DenseNet121 are used to analyse the retinopathy level of a patient. This web application can be used by clinicians, ophthalmologists and physicians. One of the major problems faced by diabetics today is the inability to track blood sugar levels that can lead to blindness. This application can also be used for diabetics diagnosis and early vision recovery.

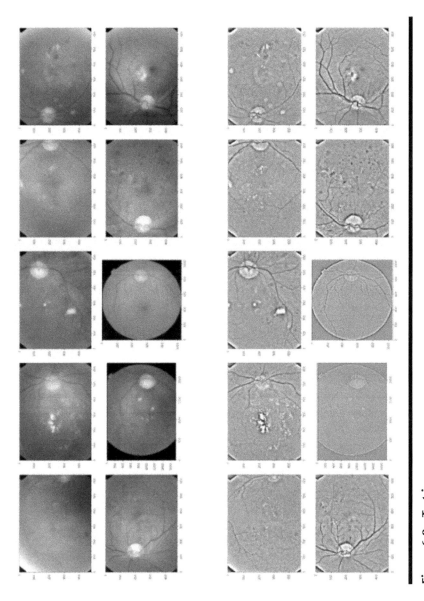

Figure 6.8 Test images.

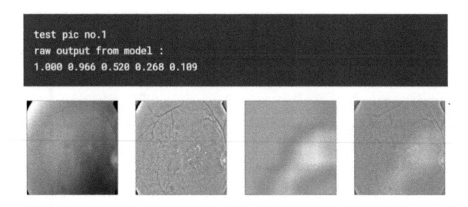

Figure 6.9 Testing images by training of DenseNet-121.

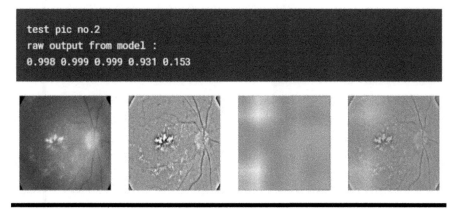

Figure 6.10   One level deeper look at DenseNet-121. Dense Block and Transition Block.

Table 6.1   Diabetics Prediction Models

Algorithm	Accuracy
Logistic Regression	71.428
K nearest neighbours	78.571
Support Vector Classifier	73.376
Naïve Bayes	71.428
Decision tree	68.181
Random Forest	75.974

Table 6.2   Diabetics Retinopathy Prediction Models

Algorithm	Accuracy
CNN	74.4
DenseNet 121	96.4

Note

1 The Diabetes Web is available from the authors at: C:/Users/Yadukrishna/Downloads/yadu_web%20(2)/yadu_web/light-bootstrap-dashboard-master/BS3/dashboard.html

References

Antal, B. and Hajdu, A. (2014) "An Ensemble-Based System for Automatic Screening of Diabetic Retinopathy". *Knowledge-Based Systems*, 60: 1–8.

APTOS (2019) "Kaggle: APTOS 2019 Blindness Detection". Available at: www.kaggle.com/c/aptos2019-blindness-detection (accessed 12 April 2021).

Bader Alazzam, M., Alassery, F. and Almulihi, A. (2021) "Identification of Diabetic Retinopathy through Machine Learning". *Mobile Information Systems*, 2021: 1–8.

California Healthcare Foundation (2015) "Kaggle Data Repository". Available at: www.kaggle.com/c/diabetic-retinopathy-detection (accessed 5 April 2021).

Chablani, M. (2017) "Towards Data Science". Available at: https://towardsdatascience.com/densenet-2810936aeebb (accessed 5 April 2021).

Conrad Stöppler, M (2021) "Definition of Diabetic Retinopathy" *RxList*. Available at: www.rxlist.com/diabetic_retinopathy/definition.htm (accessed 5 April 2021).

Douillard, A. (2018) "Densely Connected Convolutional Networks". *arXiv*. 1608 06993. Available at: https://ieeexplore.ieee.org/document/8099726

Gondhalekar, A. (2020) *Data Augmentation: Is It Really Necessary?* Delhi: Analytics Vidhya. Available at: https://medium.com/analytics-vidhya/data-augmentation-is-it-really-necessary-b3cb12ab3c3f.

Jee, G., Harshvardhan, G., M., Gourisaria, M. H., Singh, V., Rautaray, S. S. and Pandey, M. (2021) "Efficacy Determination of Various Base Networks in Single Shot Detector for Automatic Mask Localisation in a Post COVID Setup". *Journal of Experimental & Theoretical Artificial Intelligence*, 35(3): 345–364. doi: 10.1080/0952813X.2021.1960638

Narasimha-Iyer, H., Can, A., Roysam, B., et al. (2006) "Robust Detection and Classification of Longitudinal Changes in Color Retinal Fundus Images for Monitoring Diabetic Retinopathy". *IEEE Transactions on Biomedical Engineering*, 53: 1084–1098.

Niemeijer, M., Abramoff, M. D., and van Ginneken, B. (2009) "Information Fusion for Diabetic Retinopathy CAD in Digital Color Fundus Photographs", *IEEE Transactions on Medical Imaging*, 28: 775–785.

Osareh, A., Shadgar, B., and Markham, R. (2009) "A Computational Intelligence Based Approach for Detection of Exudates in Diabetic Retinopathy Images". *IEEE Transactions on Information Technology in Biomedicine*, 13 : 535–545.

Pires, R., Jelinek, H. F., Wainer, J., et al. (2013) "Assessing the Need for Referral in Automatic Diabetic Retinopathy Detection". *IEEE Transactions on Biomedical Engineering*, 60: 3391–3398.

Randive, S. N., Senapati, R. K., and Rahulkar, A. D. (2019) "A Review of Computer-Aided Recent Developments for Automatic Detection of Diabetic Retinopathy". *Journal of Medical Engineering & Technology*, 43(2): 87–99.

Roychowdhury, S., Koozekanani, D. D., and Parhi, K. K. (2014) "DREAM: Diabetic Retinopathy Analysis Using Machine Learning". *IEEE Journal of Biomedical and Health Informatics*, 18(5): 1717–1728. doi: 10.1109/JBHI.2013.2294635. PMID: 25192577.

Sil Kar, S. and Maity, S. P. (2016) "Retinal Blood Vessel Extraction Using Tunable Bandpass Filter and Fuzzy Conditional Entropy". *Computer Methods and Programs in Biomedicine*, 133 : 111–132.

Taylor, D. J., Goatman, K., Gregory, A. A., Histed, M., et al. (2009) "Image-Quality Standardization for Diabetic Retinopathy Screening". *Expert Review of Ophthalmology*, 4: 469–476.

Walter, T., Klein, J. C., Massin, P., et al. (2002) "A Contribution of Image Processing to the Diagnosis of Diabetic Retinopathy Detection of Exudates in Color Fundus Images of the Human Retina". *IEEE Transactions on Medical Imaging*, 21: 1236–1243.

Zhang, L., Li, Q., You, J., et al. (2009) "A Modified Matched Filter with Double-Sided Thresholding for Screening Proliferative Diabetic Retinopathy". *IEEE Transactions on Information Technology in Biomedicine*, 13: 528–534.

Chapter 7

IIoT Applications and Services

P. Shanmugavadivu, T. Kalaiselvi,
M. Mary Shanthi Rani, and P. Haritha

7.1 Introduction

The Internet of Things (IoT) deals with the connection of internetworked sensors, devices, machinery, and gadgets through supportive software and technologies. The IoT generates and exchanges data across the internet [1]. It is a path-breaking New Age technology that finds a place in the automation of complex and labor-intensive manual processes. It has a record of numerous applications in several domains including agriculture, healthcare, security, remote sensing, industry, weather forecasting, automated vehicles, wearable devices, and smart home appliances, IoT can be deployed in a toothbrush and can be scaled up for a massive manufacturing industry [2]. The Internet of Things variously can be found in a heart monitor that analyzes the heartbeat of a person, a biochip transponder that can provide a unique identification number for an animal [3], a sensor in an automobile that can alert the driver when the air pressure of a tire drops, etc. [4]. IoT devices can be interfaced without a keyboard or a screen. The virtual assistants such as Alexa and Siri are IoT-enabled software that gathers information across the globe. From the industrial perspective, IoT devices produce a rich set of data, which plays a vital role in automated operation control for decisive, prediction-based safety and security measures [5].

7.1.1 The Need for IoT

IoT as a transformative technology has gained popularity and momentum, within a short span of time since its introduction. This buzzword was coined in 1999 by Kevin Ashton, when he decided to track the supply chain of the Proctor & Gamble products, using Radio-Frequency Identification (RFID). The primary advantages of IoT were deemed to be the reason for its success and growth in leap and bounds.

- harnesses voluminous data for decision-making
- offers automated tracking and monitoring
- reduces workload and increases productivity
- increases efficiency
- enhances quality of life.

We will now discuss these advantages of IIOT:

- Harnesses voluminous data for decision-making: The devices or sensors connected through the IoT can be geographically apart and still can generate and collect huge volumes of data. For instance, on the domestic front, an IoT-enabled refrigerator can raise alerts on the expiry date of food items, give prompts on the food items running out of stock, and optimize power consumption. Likewise, the business-related data generated through IoT can help to strategize the business plans and solutions.
- Automates tracking, monitoring and control: Tracking enables users to check the status of the machine to take necessary actions. For example, the user can monitor quality of food items in a smart refrigerator and they can replace the items without consistently checking.
- Reduces workload and increases productivity: Since the devices are performing major tasks, this leads to a reduction in costs, time, and human intervention. Most of the communications are machine-to-machine and the connected devices lead to useful implementations that increase the efficiency of both the business and end users. For example, the lights are automatically turned off when the user leaves the smart room and this saves electricity.
- Enhances quality of life: The applications of IoT improve quality in various sectors, namely health, lifestyle of users, etc. For example, the smart watch is used to track the heartbeat rates and it can analyse the exercises carried out by the user [6].

7.1.2 Integration of IoT and Technologies

The advances in computing technologies have increased the potential of IoT, in several possible areas:

■ *Open Source Platform*: The open source software plays a major role in IoT. The Linux operating system is predominant in IoT. Arduino is a popular open source electronic platform that provides an integrated development environment (IDE) for hardware and software IoT prototypes using sensors. Devicehub.net serves as a viable interface between the interactive IoT and machine-to-machine (M2M). This software also provides support for device control, data management, and wireless control. The IoT Toolkit is an advanced platform that offers deployment services for IoT to interface with embedded agents and support prototyping with Raspberry Pi, CoAP (Constrained Application Protocol) proxies and data-compatible frameworks. SiteWhere supports the fabrication of scalable IoTs on the cloud as well as Big Data management. ThingSpeak facilitates real-time data processing and visualization. Webinos is a specialized web programming platform for platform-independent IoTs.

■ *Big Data Technologies*: The technology-driven way of life has triggered the usage of IoT and the generation of Big Data by IoT related to automobiles, healthcare, business, industry, meteorological devices, the military, smart devices, agriculture, e-governance, e-commerce, etc. As IoT is inseparable from Big Data, the implicit potential of Big Data is harnessed using the appropriate tools for data engineering, information processing and analytics, using HBase, MongoDB, etc. [7].

■ *Cybersecurity*: The IoT devices are prone to security threats due to the vulnerable open-source code, APIs (application programming interfaces), security integration, unpatched software, test cases as well as due to limited visibility on the integration of IoT components. These threats can be suitably mitigated through data encryption API security, periodic open-source software updates, authentication schemes, compliance of IoT standards and DNS (Domain Name Server) filtering [8].

■ *Software-Defined Networking (SDN)*: SDN is an emergent architecture, which offers flexible and dynamic centralized network management. It has a special feature of decoupling network control and data forwarding functions. This ensures higher-order abstraction and programmability for the network control plane [9]. Some of the SDNs available on the market are: Cisco DNA Center, Cisco ACI, IBM Cloud Internet Services, Cradlepoint NetCloud Engine, VMware NSX, and IBM Networking Services for Software Defined Networks [10].

7.2 Components of IoT

An IoT system includes four principal components: (1) IoT sensors/devices; (2) connectivity; (3) data processing; and (4) user-interface.

■ *Sensors or devices* are designed to generate or gather data from their respective environment. These gadgets are the bridge between the digital and the real world. The data can be simple, say, reading a temperature or complex, such as recording a video. An IoT set-up can have single or multiple sensors. For instance, a smartphone has multiple sensors, such as motion sensor, light sensor, geoposition sensor, environment sensor, proximity sensor, compass sensor, biometric sensor, meters, probes, gadgets, and actuators to measure various parameters.

■ *Connectivity* is the second crucial element that provides physical/ virtual connectivity among the sensors/devices through Wi-Fi, Bluetooth, cellular satellite, low-power Wide Area Networks gateway, routers, Ethernet, etc. The optimality of connectivity is decided by the quantum of power consumption, range coverage, and its bandwidth [11].

■ *Data processing* deals with handling raw data (batched or steaming data), formatting data (table, Spreadsheet, plain text, files, etc.), data storage (into local, remote or cloud storage), processing and preparing output, and visualization [12].

■ *User-interface* deals with the delivery of information to the end-user, in a human-interpretable form. This can be accomplished through command line, menu-based, form-based, voice-based, and graphical user interface (GUI). Alexa and Siri are the typical examples of voice-based interfaces.

The other important elements of IoT are security, gateway, application, and user.

In general, the seven layers of IoT are delineated as: (1) sensors; (2) sensor to gateway network (e.g., BLE, LoRaWAN, ZigBee, and Sigfox); (3) gateway (routers/ modems); (4) gateway to internetwork (Ethernet, Wi-Fi, satellite, cellular); (5) data ingestion and information processing (backend of mobile/web applications); (6) internet-to-user network (user-view of the processed data); and (7) value-added information (visualization of data for better insights leading to strategic decisions).

A few trivial applications of IoT are [13]:

■ flagging of automated notifications
■ automated alert for machine's temperature/pressure
■ real-time information monitoring real-time tracking of traffic, vehicles, geopositions, etc.
■ remote monitoring and controlling of home appliances and electronic gadgets.

7.3 Programming Software for IoT

IoT requires the support of programming languages and tools for data acquisition, data processing and manipulation, storage, and interpretation. C, C++, Java, B# and Python are prominently used for IoT programming and interfacing. As middle-level languages, C and C++ have been proved to be handy for developing interfaces and embedded software for IoTs. Python has been proved to be a simple, but powerful programming language to interface with Raspberry Pi CPU. As an interpreted language, Python has a good collection of built-in functions to handle and process volumes of data with ease.

7.3.1 Python for IoT

This section describes the exclusive supporting features of Python for IoT:

- *Raspberry Pi and Python:* Python supports I/O ports management for Raspberry Pi. The board supports Ethernet and wireless connectivity, communication and monitoring.
- *PyBoard and Python*: The PyBoard is a powerful electronics development board and it operates on MicroPython. The board is connected to a personal computer through USB (and affords a serial python prompt to make programming faster). The USB flash drive is used to store the Python scripts. MicroPython is a Python complier that readily runs on BareMetal (an exokernal-based single address space operating system) and generates the Python operating system. The pyb module contains classes and functions for controlling the board's I²C (inter-integrated circuits), UART (Universal Asynchronous Receiver/Transmitter), ADC (Analog to Digital Converter), DAC (Digital to Analog Converter), and SPI (Serial Peripheral Interface) peripherals [14].

7.3.2 Python for Backend Development

Due to its programming capabilities, Python is termed a universal programming language. Hence, Python has been proved to be handy and efficient for backend software development too.

- *Python with MQTT protocol*: MQTT (Message Queuing Telemetry Transport) is an open messaging protocol that is one of the most widely used communication mechanisms for IoT devices. Python's MQTT protocol supports high-speed data exchange with a small payload. The integration of Python and MQTT can be described as a four-step process. In the first step, the sensor chip (programmed device) collects data. The second step, the HTTP (Hypertext Transmission Protocol) request or TCP (Transmission Control Protocol)/IP (Internet Protocol/UDP (User Datagram Protocol) transmits the IoT data using a Gateway Request, This is further handled by the MQTT protocol, that receives the data from the console. Finally, the data visualization is accomplished by using matplotlib.

■ *Python-based Microsoft Azure for IoT*: Microsoft has updated its open-source IoT extension, by augmenting Azure CLI 2.0's functionalities. The Azure CLI 2.0 uses commands to communicate with the Azure Resource Manager and management endpoints. For instance, we can create an Azure Virtual Machine (VM) or IoT Hub using Azure CLI 2.0. Through the CLI extension, an Azure service can enhance Azure CLI by allowing users to gain access to exclusive services. The IoT extension gives programmers command-line access to the IoT Hub Device Provisioning Service, IoT Edge, and IoT Hub [15].

7.3.3 IDEs for Internet of Things Development (IoT)

■ *Arduino*: The Italian company Arduino fabricates microcontroller boards, interactive items, and kits, and has developed a complete IDE to interface with hardware. With a huge amount of examples and pre-loaded libraries, it is a full package. The C and C++ programming languages for programmable microcontrollers are supported by this IDE.
■ *Raspbian*: This is an exclusive IDE designed for Raspberry Pi boards. This IDE contains more than 35,000 packages and pre-compiled software that makes installation simple.
■ *PlatformIO*: This is a generic debugger in a cross-platform IDE for IoT. It contains more than 20 development platforms, over 4,000 embedded boards, and a wide variety of frameworks.
■ *EclipseKura*: This is a Java-based open source development platform for IoT apps development. It is specially designed for tasks like the Eclipse IoT challenges, monitoring industrial equipment, etc. [16].

7.4 IoT Hardware

IoT hardware encompasses a wide range of gadgets, including sensors, bridges, and routing devices. These IoT devices handle system activation, security, action definitions, communication, and the detection of support-specific objectives and actions. IoT Low-power boards and single-board processors (Arduino Uno) are tiny boards which are inserted into mainboards to improve and increase their functionalities. The GPS, light and heat sensors, or interactive displays are the common IoT hardware used. Figure 7.1 depicts one such Arduino Uno board.

The Raspberry Pi 2 is a popular IoT platform that is a small, extremely affordable computer that can function as a complete web server. It is frequently referred to as "RasPi," and it has sufficient processing and memory to run both IoT Core and Windows 10 on it. Figure 7.2 shows the Raspberry Pi and its components.

RasPi exhibits excellent processing power, when programmed with Python. A single-board computer called BeagleBoard uses an ARM (Advanced RISC Machine) processor, which is more powerful than RasPi's, that runs on a Linux-based operating system. Here, RISC stands for Reduced Instructions Set Computers.

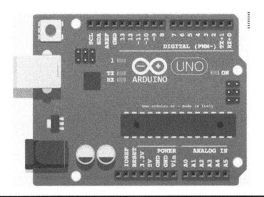

Figure 7.1   Arduino Uno.

Source: Google Images

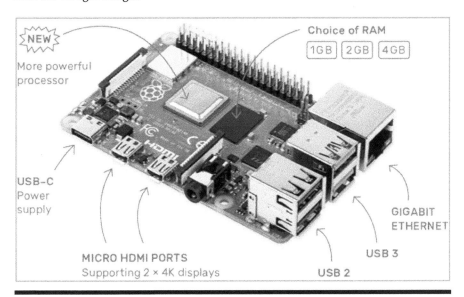

Figure 7.2   Raspberry Pi.

Source: Google Images

Galileo and Edison boards from Tech Giant Intel are further alternatives; both are suitable for larger-scale production. Qualcomm has produced a variety of enterprise-level IoT technology for everything from automobiles and cameras to healthcare [17].

The Beaglebone Black shown in Figure 7.3 has a faster processor than the Raspberry Pi (1GHz vs. 700MHz) Instead of an ARM 11-based processor. It uses an ARM Cortex A-8 processor. However, processing speed is influenced by the use cases [18].

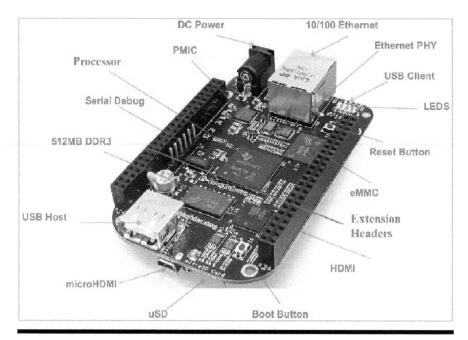

Figure 7.3   Beagle Board.

Courtesy: www.beagleboard.org

7.5 Industrial IoT (IIoT)

The existing and emerging opportunities for industrial IoT applications have propelled the usage of IoT in business, operations and manufacturing sectors in order to maximize productivity and profit as well as to reduce costs. The industrial IoT (IIoT) is a network of equipment, machinery, and sensors interconnected via the internet, in order to continuously gather and process data for better monitoring, operational, and business solutions.

7.5.1 Applications of IIoT

The scope and influence of IIoT are increasing over time. In the data-driven era, the IoT serves as one of the primary sources of data for production and business management. This section outlines the role of IoT applications and impact in the industry sector [19]:

- ■ *Equipment management and monitoring*: IIoT can remotely control many production plants across various geographical regions by using digital devices and software. The goal of gathering and utilizing the historical data is to improve the process/product and to create an environment for information-driven decision-making.

■ *Predictive maintenance*: Predictive maintenance involves identifying a machine's need for maintenance before an emergency arises and output must be stopped immediately. Thus, implementation of data acquisition, analysis, and management are needed to avoid such unexpected situations. IIoT uses sensors that can transmit alerts when specific danger indicators appear. The real-time data are gathered from the sensors on-the-fly, in order to track the robots or machines. The advanced algorithms are used to send alerts when the monitoring parameters, such as temperature, pressure, etc., cross the prescribed upper/lower threshold level.

■ *Intelligence systems for improvements:* IIoT generates vital data which is either directly or indirectly used to streamline and optimize the key resources and processes. This may lead to the creation of generic/ problem-specific operational intelligence and business intelligence models.

■ *Pinpoint inventories*: IIoT is also used for automated inventory monitoring and tracking. This process may help in optimized inventory management, supply chain management, and marketing.

■ *Quality control*: IIoT devices are deployed for quality control of process and products, at every stage of manufacturing.

7.5.2 IIoT Sensors

The sensors are the most significant component of the IoT hardware. The IoT sensors commonly used in the industry are primarily categorized into four types: (1) mechanical sensors; (2) thermodynamic sensors; (3) electrical sensors; and (4) geospatial sensors:

■ *Mechanical sensors*: The real-time mechanical IoT sensors can monitor both static (time-invariant inputs) and dynamic (time-variant inputs). These types of sensors are seamlessly used in all major industrial processes. Some of the most frequently used mechanical sensors are *gyroscopes, piezoelectric transducers, tilt sensors*, and *accelerometers*. These sensors are used by operators to gauge aircraft pitch, yaw, and weights of solids in containers and the quantity of oil in tanks.

■ *Thermodynamic sensors*: Thermodynamic sensors are designed to measure the internal energy and specific heat at a given constant pressure. Since liquids and gases are essential to several production processes, thermodynamic sensors are often used in the manufacturing industry. The most commonly used IoT thermodynamic sensors are: *temperature sensors, flow meters, gas sensors, humidity/moisture sensors, anemometers*, and *pressure sensors*.

■ *Electrical sensors*: In order to monitor the automated and manual operations involving electricity, these sensors are used. In many industrial IoT set-ups, mechanical, thermodynamic, and electrical sensors work in tandem. Electrical sensors assume importance where early/timely detection of technical faults cannot be done by human inspection. The electrical sensors are a great help in optimal utilization and safe usage of electricity. The typical electrical sensors include *current transducers, voltage sensors, power consumptions/efficiency sensors*, and *hall effect sensors*.

■ *Geospatial sensors*: Geospatial sensors monitor the physical location of objects. They can explain how assets move across an area and how machinery is positioned in relation to other industrial infrastructure. The LIDAR sensor used by many autonomous cars can track their location in relation to other objects, vehicles, and pedestrians. These sensors are used in many Google map tagged applications. Geospatial sensors are employed for surveillance since they can instantly warn security professionals about unusual movement around a secured area. Some of the geospatial sensors are: *potentiometers, proximity sensors, radar, infrared sensors*, and *photoelectric sensors* [20].

7.5.3 IoT Sensors for Industrial Automation Solutions

IoT sensors for industry include: (1) smoke sensors; (2) proximity sensors; (3) infrared sensors; (4) piezo sensors; (5) temperature sensors; (6) optical sensors; and (7) image sensors.

■ *Smoke sensors*: The incorporation of smoke sensors in IoT industrial automation systems has led to the development of several interesting use-cases. These sensors are widely used for monitoring in HVAC (heating, ventilation, and air conditioning), construction sites, and industrial units where there is a higher risk of fire and gas leaks. During accidental gas leak or fire outbreak, the smoke sensors can alert the appropriate team/personnel, thereby preventing huge loss of life or assets. Smart IoT smoke sensors are produced by Heisman and Nest Protect for both domestic and commercial applications.

■ *Proximity sensors*: These sensors calculate the distance between themselves and the nearest objects. The proximity sensors are embedded on automobile bumpers and other in-motion machinery to warn drivers/operators of impending collisions.

■ *Infrared sensors*: Infrared (IR) sensors are used to detect the presence of living beings. These sensors are widely used in military applications. In recent years, IR sensors have become a part of any industrial IoT solution. These sensors are used in the electronics, chemical, and healthcare industries. A few IoT IR sensors used in industries are *Asahi Kasei Microdevices (AKM), Murata, Melexis MLX90614*, and *Intersil ISL29021*.

■ *Piezo sensors*: The piezo sensors are used to measure pressure variations. These sensors can be used to create an industrial IoT system that can track pressure in pressure-sensitive devices including boilers, water systems, aerospace, oil-drilling systems, and many more. Some of the most popular pressure sensors used in industrial IoT applications include *Pressure Systems Series 960 and 970, the Paroscientific Inc. Series 1000, 2000 and 6000*, and *Environdata BP10 Series.*

■ *Temperature sensors:* These sensors have found application in many industrial IoT systems, such as FMCG (fast-moving consumer goods), pharmaceuticals, biotechnology, and other sectors, where temperature monitoring is essential. The common temperature sensors in industrial IoT applications include *Melexis MLX90614, Environdata TA40 Series*, and *Geokon 4700.*

■ *Optical sensors*: These sensors are designed to measure the properties of light, namely wavelength, intensity, frequency, and polarization of light waves. These sensors are capable of tracking any type of electromagnetic radiation, namely light, electricity, and magnetic field. Optical sensors are used in industrial automation process including telecommunications, elevators, and construction, healthcare and safety systems. The optical sensors *VCNL4020X01*and *TCxT1600X01* are fabricated by Vishay exclusively for industrial IoT applications.

■ *Image sensors*: In several sectors, including healthcare, transportation, and others, image sensing has immense potential to transform industrial automation. These sensors find a place in virtual monitoring in hospitals, factories, and other facilities when connected with an IoT- enabled system. The gesture analysis of a patient can send relevant messages or alerts to a physician during critical conditions. Image sensors can alert a driver to emergency situations, such as a train collision. *Omron* and *Invisage* are the successful examples of integrating image sensing into the IoT [21]. Figure 7.4 portrays the application of various sensors in operations and business.

7.6 Artificial Intelligence and IoT

Artificial Intelligence (AI) is a transformative technology that has complemented the use of IoT in improved communication, computation, analytics, and decision-making in business, as well as enhanced healthcare services. The integrated impact of AI models (machine learning and deep learning algorithms) and IoT has created new avenues to heighten the quality of human life and industrial growth, in the light of Industry 4.0.

All the types of data analytics, namely descriptive, predictive, prescriptive, diagnostic, and adaptive data analytics can be performed on IoT-generated data.

■ *Descriptive analytics*: This approach requires a huge volume of data which can be collected from the IoT systems embedded in smart devices. Such data

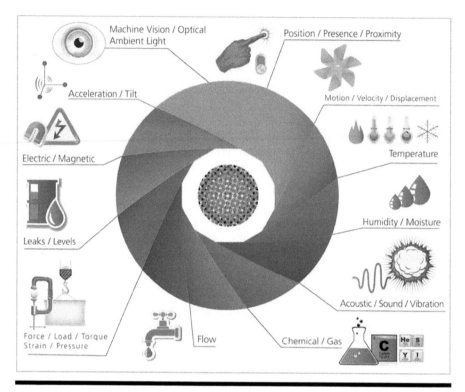

Figure 7.4   IoT Sensors.

Source: Google Images

is generally stored in the cloud environment and is managed through cloud servers and multi-core systems. This kind of analytics is conducted to explore and answer the question: "What had/has happened?"

- *Predictive analytics*: Predictive analytics uses historical or observational data and principles of advanced statistical/ probabilities and machine learning models to predict the future trends, based on the analyzed pattern. The IoT paradigm is designed to perform data acquisition from various IoT devices and can design a framework for analysis using the cloud. Some of the applications of predictive analytics are predicting supply-demand trends, prices, and maintenance. Predictive analytics proposes what and when the event will occur based on future predictions. This type of analytics provides answers to the question: "What is the future outcome?"
- *Prescriptive analytics*: Prescriptive analytics expands the capabilities of predictive analytics by adding the insights of the future forecasts to impact analysis. This type of analysis predicts future conditions; and also offers suggestions

for how the result might be used, by addressing the question: "What should be done next?" Prescriptive analytics is ideal for Industrial IoT (IIoT) settings where business intelligence-based choices are made by using the capabilities of cloud/edge computing, Big Data analytics, and machine learning.

■ *Diagnostic analytics*: This type of analytics also finds a place in IIoT applications. On identification of a faulty operation or failure of an event, diagnostic analytics finds the reasons for that occurrence. This analytics aims to address the question, "Why did it happen?".

■ *Adaptive analytics*: The results of the predictive analytics must be adjusted with real-time data during actual deployment. In order to do this, adaptive analytics are employed to modify or optimize the outcomes, based on the most recent data history and the correlation among the data.

7.7 IIoT Start-Ups in India

India is making its best effort to leverage IoT and artificial intelligence and to promote the culture of innovation on a par with the developed countries. The increased attention and sustained efforts of the Government of India, under the Make in India project, have defined new horizons of business and innovation. The promotional avenues and financial support through MSMEs (micro, small and medium-sized enterprises) have triggered the establishment of many start-ups. India has a galaxy of 1,500 IoT start-ups, which are intended to harness the potential of IoT and its applications perspectives. This section highlights a few such IoT start-ups.

■ *Stellapps Technologies*: Stellapps is one of the leading IIoT firms in India that has gained enormous growth by offering top-notch services. This Bangalore-based company is primarily focused on data collection and machine learning. The SmartMooIoT platform of Stellapps collects data, using sensors built into animal wearables, milking systems, milk chilling machinery, and milk procurement accessories. The gathered data is sent to the Stellapps SmartMoo™ Big Data Cloud Service Delivery Platform (SDP), wherein the StellappsSmartMoo™ suite of applications extract and analyse the information before sending the results of the analytics and data science to various stakeholders via low-end and sophisticated mobile devices [22] (www.stellapps.com).

■ *Parentheses Systems PLC*: Parentheses Systems, a deep-tech start-up, is thriving on shifting the Industrial IOT ecosystem from a technology-driven ecosystem to a behavior-led transformation for small and medium-sized discrete manufacturing companies. The HuMaC (Human Machine Convergence) platform allows the orchestration of the "Data-Decision" cycle to maximize man-machine margins. HuMaC is a human-centric paradigm of artificial intelligence and people working together harmoniously to enhance both new experiences and cognitive abilities such as decision-making and gaining knowledge from business insights. While artificial intelligence creates robots that behave and perform in a similar way to people, augmented intelligence makes use of those same devices to improve the output of human work and the Human Machine Ecosystem [23] (www.parentheses.systems).

■ *MLWorkX:* MLWorkX is a manufacturing supply chain company that focuses on end-to-end supply chain solutions for custom manufacturing. The company is an IoT-powered cloud-based manufacturing platform, offering seamless and optimized manufacturing services by making custom product manufacturing as simple as ordering on Amazon. The company helps their clients to reach the market faster by transforming product ideas into physical manifestations with complete online visibility (www.mlwo rkx.com).

■ *Utvyakta:* This company works on preventative maintenance and energy-saving technology exclusively designed for air compressors. They offer analytics for deriving performance and utilization insights as well as monitoring of crucial metrics. Any ERP (enterprise resource planning)/data management IoT layer available on the factory floor or shop floor can be integrated with it. This company collaborated with several of the world's top air compressor producers as well as the compressor users in the oil and gas, automotive, mining and construction, and cement industries. Additionally, the company offers retrofitting for already-built installations. In addition to actual cost reductions, they offer a 20 percent increase in overall efficiency of on-site air compressors, assuring cost saving and improved performance (https://utvyakta.com).

■ *Smelco:* Smelco is an IoT-based anti-theft shutter infrared sensor security system that guards the store against any robbery. One can watch over the store from anywhere in the world. The user can insert a SIM card into the device, even without Wi-Fi. When someone tries to pull the shutter, the SIM card is activated and the customer can receive alerting alarms. In addition, a motion sensor is provided with dual security to prevent backdoor entry (https://sme lco.in).

■ *Galan-fi Smart IoT Technology Pvt Ltd:* Galan-Fi is an IoT-powered predictive analytics model that offers insights to save energy, gas, water, and other utilities. This company was founded by women entrepreneurs in 2021. They manufacture IoT-based energy and asset management systems (www.zaubac orp.com).

■ *Zenatix:* Zenatix offers amazing energy-saving solutions for large commercial electricity consumers through machine learning models. It offers business including Wattman and Wattman Lite, which contributes to a 30 percent reduction in electricity use [24] (www.zenatix.com).

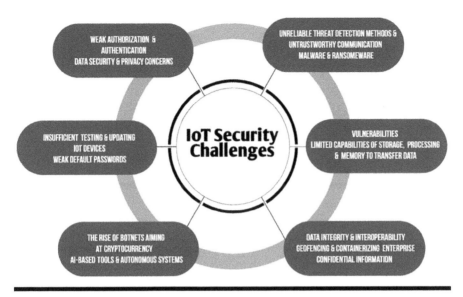

Figure 7.5   IoT security challenges.

7.8 Challenges in Securing IoT in India

IoT is a combination of hardware and software, so it is quite vulnerable to security threats. The entire IoT network functions in a single device and the security of every other device is put at risk, when the master device has a fault. A few manufacturers in the IoT market address security concerns of IoT devices and their data access. The Data Security Council of India (DSCI) has enumerated the common challenges of IoT security, which offer newer avenues for business and research (Figure 7.5).

TechSagar manages the business and research capabilities of various entities from the IT industry, start-ups, academia, and research & development institutes. TechSagar is India's cybertech repository, which is supported by the office of National Cyber Security Co-ordinator and managed by DSCI that provides a platform to discover India's cybertech capabilities. Additionally, TechSagar reports that there are more than 700 businesses, 129 academic institutions, 22 research and development centers, and 243 researchers working in the IoT field in India. The IoT capabilities described by TechSagar are depicted in Figure 7.6.

7.9 Policies and Regulations for Promoting IoT in India

■ *Draft IoT Policy 2015*: The initial draft of India's IoT Policy Document was released by the Ministry of Electronics and Information Technology (MeitY,

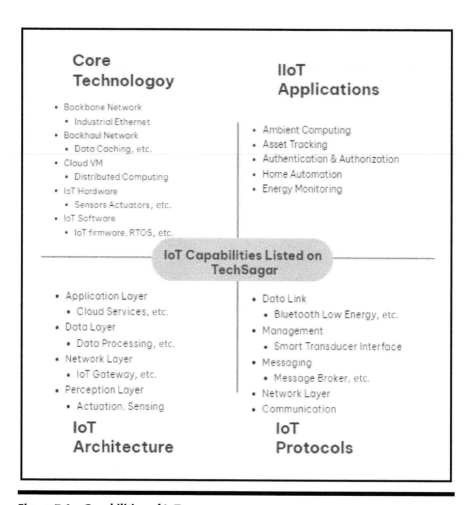

Figure 7.6   Capabilities of IoT.

formerly known as DeitY) in 2015, which is a robust governance structure for the broad adoption and execution of IoT-related regulations.

■ *National Digital Communications Policy (NDCP)*: The NDCP has innovative objectives and policy measures to solve the issue of communications and access to digital services in India.

■ *Smart City Mission*: The goal of the Smart City Mission is to use technology to produce better outcomes in order to boost economic growth and enhance quality of life.

■ *IoT Centre of Excellence (CoE) by NASSCOM, MeitY and ERNET*: The greatest deep tech innovation ecosystem in India is CoE. It was created exclusively to assist Indian IoT start-ups to produce market-leading goods.

■ *IoT lab, a partnership between IIT-Delhi and Samsung*: A Memorandum of Understanding (MoU) was signed between Samsung and IIT-Delhi in 2016 with the aim of enhancing research capabilities and industry collaboration in the IoT domain.

7.9.1 Recommendations for IoT Devices

These IoT-based smart gadgets, such as wireless applications, smart appliances, home security sensors, etc., possess a great opportunity for hackers to discover vulnerabilities or gaps. Therefore, it's crucial to ensure that there is no room for cyber risks. In India, there is increasing emphasis on intervention through research. The following suggestions underscore the latest developments in this domain.

A powerful cryptographic method is needed to safeguard our sensitive data generated by IoT. The standard encryption techniques, however, do not work with the restricted IoT hardware. Currently, a fresh area of cryptography is under development to fix this problem. The primary objectives of cryptography are to achieve confidentiality, integrity, authenticity, and availability. Lightweight cryptography strives to deliver crypto-solutions customized to restricted contexts.

According to TechSagar's IoT research, the government could aid the private sector in creating long-term business plans for 5G-enabled IoT services by creating an ecosystem of applications, laying out the necessary network infrastructure, and forging strategic alliances.

Additionally, it suggests that the government and major industry participants should acknowledge and reward IoT start-ups and solution providers as well as encourage talent development.

In order to understand and develop IoT security, prime areas like frameworks, standards, collaboration efforts (between DoT, NCIIPC, NCSC, MeitY, etc.), it is essential to invest in skilling, capability-building through training, product testing, and certification [25].

7.10 Applications of IoT

IoT has a wide spectrum of applications in the primary sectors such as healthcare, agriculture, security, embedded systems, robotics and much more.

■ *Smart Homes*: The integration of IoT intelligent utility systems and entertainment together forms Smart Homes. The utilities in a Smart Home include electricity meter, locking systems, television, the lighting systems, surveillance systems, refrigerator, washing machine, etc., with IoT to monitor the respective activities.

■ *Smart City*: Cities use IoT-enabled Information and Communication Technologies (ICT) to manage traffic flows, electricity, water distribution, transport, resources, and public services. Smart city mission aims at improving the economic growth and increasing the well-being of the citizens [26].

■ *Smart Farming*: IoT, drones and ICT play a very important role in managing agricultural practices, monitoring the related parameters such as micro-climatic conditions, soil moisture and nutrients, irrigation and supply of fertilizers and pesticide. These technologies enable the farmers to increase the productivity with minimized resources, cost, and efforts [27].

■ *Retail Business*: Sales data from both online and offline retail can manage warehouse robotics and automation using data from IoT sensors. The analytics on IoT data helps in strategic planning and decision-making, aiming at profit maximization, market expansion and penetration, product modification, etc.

■ *Wearable devices*: The fitness bands and smart watches track a variety of health-related parameters, including calories burned, distance traveled, heart rate, and blood oxygen saturation. There are many other specialized health care IoT gadgets that connect the patients and doctors round the clock.

■ *Traffic monitoring*: IoT plays a major role in smart city planning. For instance, it helps metropolitan cities to manage their automobile traffic flows, using Google Maps or Waze to collect and exchange data from cars using mobile phones as sensors. In addition to aiding in traffic monitoring, it provides information on the various traffic conditions, expected arrival times, and distances to destinations.

■ *Fleet management*: The installation of IoT sensors in fleet cars establishes effective communication between the drivers, vehicles, and management. This guarantees that owners and drivers are well-informed on the condition, functionality, and requirements of vehicles on the move.

■ *Water supply*: A sensor coupled with software that is externally placed in water meters aids in the seamless data collection, processing, and analysis of data, allowing for better knowledge about consumers' behavior, supply service issues detection, and results reporting [28].

7.11 Use-Cases in Industrial IoT

7.11.1 Predictive Maintenance

Periodic maintenance of assets and resources apparently pays off in millions of dollars of operational cost saved, in any business. Sensors, cameras, and data analytics help managers from various industries to accurately predict the likely time of breakdown before it really occurs. These IoT-enabled systems can generate warning signs and use such information to develop maintenance schedules, as well as perform preventative maintenance on equipment before issues can arise. Thus, IoT-based sensors transform the maintenance process into something dynamic and automatic.

The right information at the right time by the IoT devices helps the managers to receive the relevant information related to the maintenance, so that the mitigating tasks can be planned well ahead. The predictive maintenance benefits in increasing safety and extending the lifetime of the plants/equipment as well as avoiding accidents and hazards [29].

7.11.2 Smart Metering

A smart meter is an internet-enabled device that detects the amount of energy, water, or natural gas used by a building or residence. The conventional meters can track only the total consumption whereas smart meters are able to detect the quantum of resource consumption, along with the time-stamp. Industries make use of smart meters to track the consumer usage as well as to fix the price as per the usage. Moreover, this approach is helpful in hazardous industries like mining as well as those with widely dispersed equipment or heavy reliance on environmental conditions. To avoid squandering resources or overloading its energy-generating equipment, the energy sector relies on smart metering and remote monitoring [30].

7.11.3 Location Tracking

GPS systems, RFID tags, and other wireless technologies can be used to track assets and locations and provide up-to-date information to the companies.

7.11.4 Location Services

These are frequently used by logistics organizations to track goods whilst in transit; and to reroute the drivers, when needed. Location-based IIoT technology can be used in industrial warehouses to track the materials. These services enable the employees to quickly locate the materials.

7.11.5 Remote Quality Monitoring

IoT sensors are used by organizations such as the Environmental Protection Agency to continuously monitor the quality of environmental resources and goods, using water sensors, air quality sensors, etc., based on their timely, reliable, relevant, and accurate data on environmental pollutants. These IoT sensors can be used by industries related to chemical processing and pharmaceutical to remotely check the quality of materials or products. Employees can quickly verify numerous processes, thus the remote monitoring tasks increase productivity. Similar to this, real-time notifications promote quicker reactions, preventing accidents that, if not caught in time, could endanger the product [31].

7.11.6 Supply Chain Management and Optimization

Supply chain management and optimization processes unquestionably use IoT-generated data. Suppliers can easily keep track of where the consignments are; where they have been stranded; and find the predicted timeline to reach a certain place. Product locations can be predicted and verified using GPS and other associated devices and sensors. This lets the customers know the actual day and time related to the shipment. The primary areas that IoT addresses in supply chain management and optimization are: authentication of location using sensors in a container, tracking traffic flow and product movement speed, monitoring; maintaining the quality parameters of the products, and optimal route detection [32].

7.12 Future of IIoT

It is estimated that the use of IIoT-enabled devices will double in future. The transformative and destructive technologies namely AI, cloud storage and computing, edge computing, mobile computing robotics, augmented reality, virtual reality, human-computer interface (HCI), etc., will expand the scope of IIoT and trigger the design and development of innovative IoT applications and products [33].

7.13 Conclusion

This chapter has dealt with the basics of IoT, IIoT, hardware and software support for IoT, types of IIoT sensors, security concerns, applications, domains and the future scope of IIoT. This technology is expected to record an exponential growth due its anticipated expansion of application domain and innovations.

References

1. www.oracle.com/in/internet-of-things/what-is-iot/
2. https://aws.amazon.com/what-is/iot/
3. www.techtarget.com/iotagenda/definition/Internet-of-Things-IoT
4. www.scottgroup.ie/internet-of-things-iot
5. www.networkworld.com/article/3207535/what-is-iot-the-internet-of-things-explained.html
6. https://medium.com/zeux/the-internet-of-things-iot-5-reasons-why-the-world-needs-it-125fe71195cc
7. www.channelfutures.com/security/why-is-iot-popular-because-of-open-source-big-data-security-and-sdn
8. www.comptia.org/content/articles/what-is-iot-cybersecurity#:~:text=IoT%20security%20is%20one%20of,attacks%2C%20such%20as%20through%20botnets

9. www.ciena.com/insights/what-is/What-Is-SDN.html
10. www.trustradius.com/software-defined-networking
11. www.leverege.com/iot-ebook/how-iot-systems-work
12. https://trackinno.com/iot/how-iot-works-part-3-data-processing/
13. www.biz4intellia.com/blog/7-layers-of-iot-what-makes-an-iot-solution-comprehensive/
14. https://trackinno.com/iot/how-iot-works-part-4-user-interface/
15. www.javatpoint.com/internet-of-things-with-python
16. www.topcoder.com/thrive/articles/python-in-iot-internet-of-things
17. www.it4nextgen.com/best-development-software-ide-internet-things-iot/
18. https://data-flair.training/blogs/iot-hardware/
19. https://nexusintegra.io/7-industrial-iot-applications/
20. https://blog.wellaware.us/blog/ultimate-guide-types-of-iot-sensors-transmitters-industrial-applications
21. www.embitel.com/blog/embedded-blog/7-most-commonly-used-sensors-for-developing-industrial-iot-solutions
22. www.bisinfotech.com/top-iiot-companies-in-india-industry-leaders/
23. www.f6s.com/companies/industrial-iot/india/co
24. https://startuptalky.com/iot-startups-india/
25. www.dsci.in/blogs/iot-technology-in-india/
26. www.jigsawacademy.com/top-uses-of-iot/
27. www.iotsworldcongress.com/iot-transforming-the-future-of-agriculture/
28. www.simplilearn.com/iot-applications-article
29. www.ibm.com/blogs/internet-of-things/top-5-industrial-iot-use-cases/
30. www.datamation.com/trends/industrial-iiot-use-cases/
31. www.techtarget.com/iotagenda/feature/Top-5-industrial-IoT-use-cases
32. www.mytechmag.com/iot-in-manufacturing/
33. https://doi.org/10.1007/978-3-030-37468-6_27

Chapter 8

Design of Machine Learning Model for Health Care Index during COVID-19

Nishu Gupta, Soumya Chuabey, Ity Patni, Vaibhav Bhatnagar, and Ramesh Chandra Poonia

8.1 Introduction

The lockdown due to COVID-19 very badly affected the economies of almost every countries but its degree of impact will be based on various factors like the spread of the virus, the duration of covering all the stages of coronavirus and the recovery rate of the patients. Slowdown in the economy in the years 2017–2018, 2018–2019 and now 2019–2020 have made it difficult to achieve the target of $5 trillion economy by 2024–2025. India lost its fifth position in the GDP rankings in 2018 compared to 2017. The consumption of consumer goods has slowed down, which has affected the share of capital investment in India. With a slump in production and consumption, the dream of being the third largest economy seems far away. In 2017, the size of the Indian economy stood at $2.65 trillion, the fifth largest economy. In 2018, the growth rate of India in comparison to the United Kingdom and France was minimal as the former grew by 3.01 percent to $2.73 trillion and the latter economies saw a growth of 6.8 percent and 7.3 percent, respectively, due to which India lost its fifth position and was pushed into seventh position in the World Bank's GDP rankings in 2018 [1]. The pandemic has affected all sectors due to the complete lockdown, and the health care sector, which was assumed would remain unaffected, is also no exception. This chapter is an attempt to observe the volatility in the Health Care Index of the Bombay Stock Exchange (BSE) and to predict the future movement of the Index using the Auto Regressive Integrated Moving Average (ARIMA) model. The Health Care Index of the BSE reported the biggest monthly gain in 21 years in the month of April 2020. The index rallied around 27 percent in the same month [2]. The hope of investors that demand would be created is connected to the development of a vaccine against this deadly disease.

The Indian health care sector can be a valuable bet as it includes pharma stocks too. India has done remarkably well in the field of generic medicines in the pharmaceutical industry and is the largest supplier of generic medicines globally. The Indian pharmaceutical sector industry accounts for over 50 percent of global demand for various vaccines, nearly 40 percent of demand from the USA and 25 percent from the United Kingdom are fulfilled by the Indian pharmaceutical manufacturers and suppliers. India supports the second largest share of pharmaceutical and biotech workforce on the global platform. The valuation of the pharmaceutical industry in India was calculated at $33 billion in 2017 [2] and $38 billion in June 2019 [3]. The turnover of India's domestic pharmaceutical market in 2018 was measured as Rs 1,29,015 ($18.12 billion), which grew at 9.4 percent year-on-year from Rs 1,16,389 crore ($17.87 billion) in 2017 [1]. India has earned a very good reputation in the pharmaceutical industry at a global level by developing a large pool of engineers and scientists who have been instrumental in winning appreciation for their research work, and they have the potential to direct the industry forward to achieve more accolades at global level., Around 80 percent of the antiretroviral drugs which are used to fight AIDS (Acquired Immune Deficiency Syndrome), a disease which has killed millions of people across the globe, are supplied by Indian pharmaceutical manufacturers and suppliers. The pharmaceutical and allied sectors like health care and insurance have contributed in a major way to India's economic growth and have played a very important role in the comparatively lower mortality rate in India in recent years. Estimates suggest that this industry has become a major source of employment creation in India with over 2.7 million people dependent directly or indirectly on highly skilled areas like manufacturing and research & development activities. The industry is among the top five best performing sectors which generates a trade surplus of over $11 billion and it has helped the Indian economy to reduce the burden of trade deficit [4]. Foreign direct investors have shown tremendous interest in the Indian pharmaceutical industry and they have invested more than $2 billion over the past three financial years, making it one of the top 10 priority sectors attracting foreign direct investment (FDI).

The Indian government is focusing more upon personal hygiene and care and has significantly increased the Personal Hygiene Cost (PHC) in the short run. It will help in creating community awareness of hygiene, which will have a positive impact on health care and allied industries like the insurance sector and the pharmaceutical sector, not only in India but at a global level. Due to the nationwide lockdown, the general medical facility has been badly impacted and has created a gap in the care of patients with other health problems, especially chronic diseases. This will have a long-term impact on the health care industry as the patients' situation will become more critical as their treatment was postponed.

The aims of this chapter are:

- to design a time series model for health care closing price data, starting from 1 February 1999 to March 2020.
- to study the impact of COVID-19, by comparing actual closing prices and forecast values from the developed model, starting from 23 March 2020 to July 2020.
- to forecast the trend up to December 2020 based on the model developed.

8.2 Literature Review

Prediction of the price of a stock or index has always spurred the interest of investors. [5] revealed that ARIMA is useful for predicting short-term movement in prices. [6] summarized the performance through six forecasting methods, such as minimum error, maximum error, mean error, standard deviation, root mean square error and ARIMA, the forecast accuracy was very acceptable. [7, 8] discussed technical, fundamental long- and short-term approaches for prediction and highlighted that a combination of robust tests would provide accurate forecasting. [9, 10] fitted the ARIMA model for 10 years and 2 years RMSE and MEA observed for accuracy prediction. This shows that short-term prediction was more accurate compared to long-term prediction. [11] applied the model to 56 different stocks in various sectors. The prediction percentage of model fits at 85 percent. [12] attempted the model fit on NSE-Nifty Midcap 50 companies and the trend for the future is observed. [13, 14] also attempted to forecast the future unobserved prices of the index and data validated with Sensex movements of 2013. [15] explored how the German market behaved in the same manner as North America, and the Asian and European markets responded.

8.3 Time Series Data

To predict the future stock price, many machine learning models have been developed, and Auto Regressive Integrated Moving Average (ARIMA) is one of them. Time series data are used in ARIMA which is also known as the Box-Jenkins Methodology [16, 17]. This is one of the most appropriate models for forecasting financial data for the short term. The following steps are followed to predict the future:

■ *Stationarity*: To forecast accurately, stationarity of the data set is required [18]. Stationarity of the data set means the behavior of the series remains same all the time.

■ *Testing for stationarity*: Stationarity is checked by applying the unit root test and is somewhat illogical in real phenomenon. Converting the data set into stationarity does not mean that the effect of the shock dies.

■ *Plotting the data*: First, plotting of the data set is done to know what differencing is required to convert the data into stationary form.

■ *Unit root test*: The Augmented Dickey-Fuller (ADF) and Philip-Perron Test are performed to know the stationarity of the time series data.

■ *Differencing of Time Series*: To convert non-stationery series into the stationary series, the first or second differencing is performed:

$$\Delta R_t = R_t - R_{t-1} \tag{8.1}$$

$$\Delta R_t = R_t - BR_t \tag{8.2}$$

$$\Delta R_t = (1 - B)R_t \tag{8.3}$$

■ The stationarity of the series is obtained through the dth difference of R_t.

$$\Delta_t^d R = (1 - B)_t^d R \tag{8.4}$$

■ *Univariate models*: Under the univariate model, only one variable is measured over time based on past series. The Box-Jenkins is developed to explain the process of generating series as well as of forecasting. Further, ARIMA is used for analysis purposes.

■ *ARIMA model*: A P_{th}-order and q_{th}-order autoregressive model is integrated with moving average.
The general equation for the ARIMA model is:

$$R_t = \varnothing_1 R_{t-1} \ldots \ldots \ldots \varnothing_2 R_{t-p} + \varepsilon_t + \theta_1 \varepsilon + \theta_2 \varepsilon_{t-2} + \ldots \theta_q \varepsilon_{t-q} \tag{8.5}$$

Where:

R_t = the actual data set used for converting the non-stationery data into stationery data by differentiating the time series value.

$\varnothing, \theta$ = Parameters

P = The time series value that are regressed on their own lagged values

q = Moving average value that is number of lagged values of the error term

ϵ_t = The random error at t

The procedure for processing ARIMA model is done in a few steps, namely testing, identification, estimating and forecasting [19, 5].

■ *Model fit checking*: Model fit checking needs to be performed to know the reliability and validity of data. Further, prediction of future pattern of series can be performed.

8.4 Development of the ARIMA Model

The following steps are considered to develop the ARIMA model starting from February 1999 to December 2020:

■ *Linear trend model*: Trend analysis of the closing price is done using a linear trend model. The fitted trend equation and accuracy measure are shown in Table 8.1. The graph is shown in Figure 8.1.

$$Yt = -16672 + 3.21555t \tag{8.6}$$

■ *Test of stationarity*: A stationary time series is one whose statistical properties such as mean, variance, autocorrelation, etc. are all constant over time. Most statistical forecasting methods assume that the time series can be rendered approximately stationary (i.e., "stationarized") through the use of mathematical transformations. A stationarized series is relatively easy to predict: you simply predict that its statistical properties will be the same in the future as they have been in the past. The Augmented Dickey-Fuller Test is applied to test the stationarity of the data. The hypothesis of the ADF test is:

✓ Null hypothesis of ADF test (H_0): Data is not stationary.
✓ Alternate hypothesis (H_A): Data is stationary.

Table 8.1 Accuracy Measures of the Linear Trend Model

Accuracy measure	Value
MAPI	41
MAD	1665
MSD	34497856

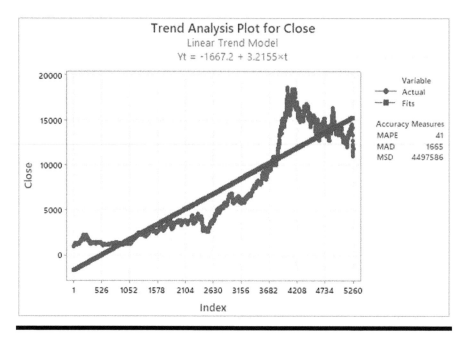

Figure 8.1 Closing prices (pattern).

Table 8.2 Augmented Dickey-Fuller Test

Augmented Dickey-Fuller Test	
Dickey-Fuller	-2.5247
Lag order	4
p-value	.35

In Table 8.2, since the p value is greater than 0.05, it can be inferred that the given dataset is not stationary.

The first difference linear equation model: First differencing is performed to convert the data into stationary form. Figure 8.2 shows the same behavior after the first differencing. The first difference linear equation model is shown below:

$$Yt = 2.37 - 0.000097 \times t \tag{8.7}$$

The accuracy measure of the first difference linear model is shown in Table 8.3.

■ *Auto correlation function and partial auto correlation function:* The coefficient of the correlation between two values in a time series is called the autocorrelation function (ACF). For example, the ACF for a time series ytyt is given by:

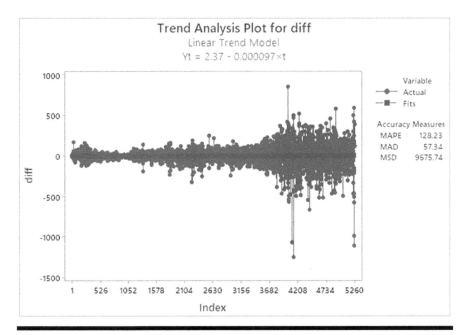

Figure 8.2 Graph after first differencing.

Table 8.3 Accuracy Measure of First Difference Linear Model

Accuracy measure	Value
MAPE	128.23
MAD	57.34
MSD	9675.74

$$\text{Corr}(yt,yt-k),k=1,2,\ldots.\text{Corr}(yt,yt-k),k=1,2,\ldots.$$

This value of k is the time gap being considered and is called the lag. A lag 1 autocorrelation (i.e., k = 1 in the above) is the correlation between values that are one time apart. More generally, a lag k autocorrelation is the correlation between values that are k time periods apart. The graph of ACF is shown in Figure 8.3.

The ACF is a way to measure the linear relationship between an observation at time *t* and the observations at previous times. If we assume an AR(*k*) model, then we may wish to only measure the association between ytyt and yt–kyt–k and filter out the linear influence of the random variables that lie in between (i.e., yt–1,yt–2,... ,yt–(k–1)yt–1,yt–2,...,yt–(k–1)), which requires a transformation of the time series. Then by calculating the correlation of the transformed time series we obtain the partial autocorrelation function (PACF). The graph of PACF is shown in Figure 8.4.

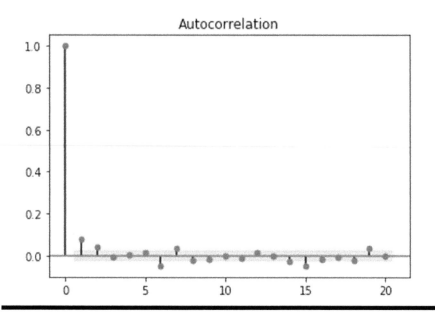

Figure 8.3 Auto correlation function (ACF).

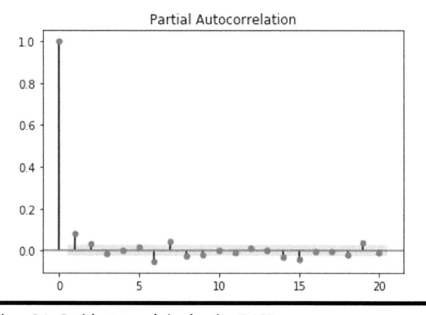

Figure 8.4 Partial auto correlation function (P ACF).

Table 8.4 Values of p, d and q

S.N.	PDQ	Error	AIC
1	000	250	63240
2	100	249	63208
3	010	415	66428
4	001	249	63217
5	110	346	65057
6	011	251	63239
7	101	249	63205
8	**111**	**249**	**63207**
9	200	249	63202
10	201	249	63204

■ *Auto correlation function and partial auto correlation function:* In this stage, the ARIMA model is applied with different values of p, d, q which are shown in Table 8.4. MSE is found between the forecasting values starting from 1 March 2020 to 31 August 2020.

Since we have seen from Figure 8.4 and Table 8.4, we have selected the model 111 for further forecasting.

■ *Forecasting:* The forecast is from 1 February 19999 to 30 August 2020 for February 2021. The Health Care Index follows the zigzag pattern which signifies the high risk. Investors should take due precautions and do the proper analysis before investing their hard earned money in this sector.

1. Using trend analysis:

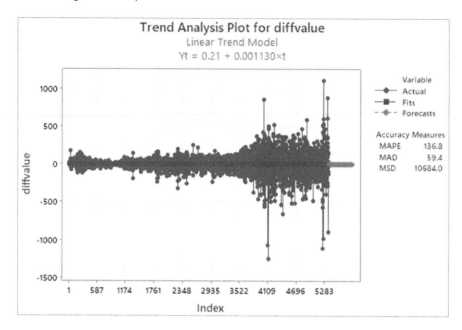

2. USING ARIMA (1,1,1)

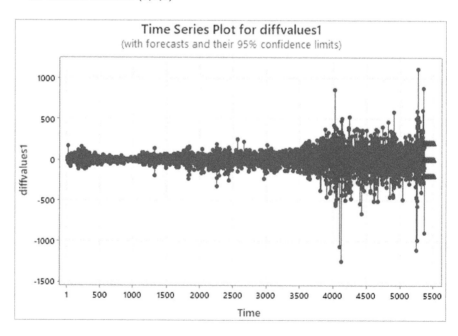

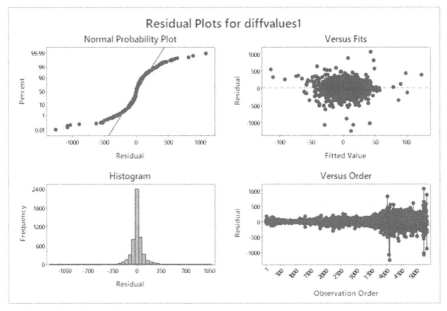

Both graphs values show constant values.

8.5 Conclusion

The research has been undertaken to forecast the closing price of the BSE Health Care Index through the ARIMA model and to show the effect of COVID-19 on the closing price which is decently predicted by the ARIMA model. The ARIMA model should be used for short-term forecasting and it also helps the investor in making the short-term investment predictions. It is suggested that investors do not panic by investing in health stocks as many things are postponed, such as operations, elective surgeries, major treatments, international patients, etc. till life get back to normal. Thus, this sector is showing the same pattern that is existing nowadays that will be followed in the future also, as suggested by the data.

References

1. Radhakrishnan Vignesh and Sumant Sen (2019) "The Downturn in the Indian Economy Continues." *The Hindu*, 6 August. Available at: www.thehindu.com/data/the-downturn-in-the-indian-economy-continues/article28836274.ece.
2. IBEF (n.d.) "Brand India." Available at: www.ibef.org/industry/indian-pharmaceuticals-industry-analysis-presentation.
3. Indian Pharmaceutical Alliance (2019) *The Indian Pharmaceutical Industry -The Way Forward*, Indian Pharmaceutical Alliance, June, pp.1–33.

4. FICCI (2013) "Indian Life Sciences: Vision 2030 Expanding Global Relevance and Driving Domestic Access." 2013-14.1-7.

5. Adebiyi A. Ariyo, Adewumi O. Adewumi, and Charles K. Ayo (2014) "Stock price Prediction Using the ARIMA Model." Paper presented at 2014 UKSim-AMSS 16th International Conference on Computer Modelling and Simulation. IEEE.

6. Jaydip Sen, and Tamal Chaudhuri (2017) "A Time Series Analysis-Based Forecasting Framework or the Indian Healthcare Sector." *Journal of Insurance and Financial Management*, 3(1).

7. Dev Shah, Haruna Isah, and Farhana Zulkernine "Stock Market Analysis: A Review and Taxonomy of Prediction Techniques." *International Journal of Financial Studies*, 7(2): 26.

8. Sheikh Mohammad Idrees, M. Afshar Alam, and Parul Agarwal (2019) "A Prediction Approach for Stock Market Volatility Based on Time Series Data." *IEEE Access*, 7: 17287–17298.

9. Madhavi Latha Challa, Venkataramanaiah Malepati, Siva Nageswara and Rao Kolusu (2018) "Forecasting Risk Using Auto Regressive Integrated Moving Average Approach: An Evidence from S&P BSE Sensex." *Financial Innovation*, 4(1): 24.

10. George E. P. Box, et al. (2015) *Time Series Analysis: Forecasting and Control.* Hoboken, NJ: John Wiley & Sons.

11. Prapanna Mondal, Labani Shit, and Saptarsi Goswami (2014) "Study of Effectiveness of Time Series Modeling (ARIMA) in Forecasting Stock Prices." *International Journal of Computer Science, Engineering and Applications*, 4(2): 13.

12. B. Uma Devi, D. Sundar, and P. Alli (2013) "An Effective Time Series Analysis for Stock Trend Prediction Using ARIMA Model for Nifty Midcap-50." *International Journal of Data Mining & Knowledge Management Process*, 3(1): 65.

13. Debadrita Banerjee (2014) "Forecasting of Indian Stock Market Using Time-Series ARIMA model." Paper presented at 2nd International Conference on Business and Information Management (ICBIM). IEEE.

14. C. Narendra Babu and B. Eswara Reddy (2014) "Selected Indian Stock Predictions Using a Hybrid ARIMA-GARCH Model." Paper presented at 2014 International Conference on Advances in Electronics Computers and Communications. IEEE.

15. Jeffrey E. Jarrett, and Janne Schilling (2008) "Daily Variation and Predicting Stock Market Returns for the Frankfurter Börse (Stock Market)." *Journal of Business Economics and Management*, 9(3): 189–198.

16. Pier Paolo Ippolito (2019) "Stock Market Analysis Using ARIMA." May 21. Available at: https://towardsdatascience.com/stock-market-analysis-using-arima-8731ded2447a.

17. Andhra, M. Pradesh, Venkataramanaiah, and Golden Valley Integrated Campus. (2018) "Forecasting Time Series Stock Returns Using ARIMA: Evidence from S&P BSE Sensex." *International Journal of Pure and Applied Mathematics*, 118: 24

18. J. Van Greunen, A. Heymans, C. Van Heerden, and G. Van Vuuren (2014) "The Prominence Of Stationarity in Time Series Forecasting." *Studies in Economics and Econometrics*, 38(1): 1–16.

19. Milind Paradkar (2017) "Forecasting Stock Returns Using the ARIMA Model." Quant Institution, Blog, March 9. Available at: https://blog.quantinsti.com/forecasting-stock-returns-using-arima-model/>.

Chapter 9

Ubiquitous Computing and Augmented Reality in HCI

Nancy Jasmine Goldena and Thangapriya

9.1 Introduction

Human-computer interaction (HCI) is a branch of computer science that studies how people and computers interact. HCI researchers study how people interact with computers and create technologies that enable people to connect with computers in unique ways. HCI is at the crossroads of computer science, cognitive science, design, mass communication, and a number of other specialties. The most fundamental and essential aspects of HCI are augmented reality (AR) and ubiquitous computing (UC) [1–7].

9.2 Ubiquitous Computing (UC)

UC is a new computing model characterized by the seamless integration of hundreds of thousands of self-communicating small-scale computers and intelligent devices found in the user's environment and daily activities. UC is also known as pervasive computing, everyware, and ambient intelligence. UC refers to the emerging trend of embedding computing capabilities, particularly in the form of microprocessors, into everyday objects in order to enable them to efficiently transmit information. UC can perform useful functions while reducing end-user interaction with computers. For

DOI: 10.1201/9781003424550-9

example, a user may want to drive from home to the gym. Smartphones, wearable devices, and other fitness trackers may be required. Wireless communication and networking technologies, mobile devices, embedded systems, wearable computers, radio frequency identification (RFID) tags, middleware, and software agents are all examples of UC [8].

9.2.1 UC's History

Mark Weiser, the Chief Technologist at Xerox Palo Alto Research Center (PARC), coined the term "ubiquitous computing" in 1988. Weiser published some of the initial papers on the subject, both alone and with PARC Director and Chief Scientist John Seely Brown, primarily defining it and sketching out its significant obstacles. The Active Badge, a "clip-on computer" the size of an employee ID card, was designed at the Olivetti Research Laboratory in Cambridge, England, allowing businesses to track the exact locations of people in a building as well as the objects to which they were attached [8].

At PARC, Weiser was inspired by social scientists, anthropologists, and philosophers, and he explored a new approach to computers and connectivity. He believes that people live by following their routines and knowledge bases, so that the most powerful things are effectively invisible in usage. This is a problem that affects the entire field of computer science. First and foremost, it makes the world a better place by activating computers. Hundreds of wireless computer devices of different sizes, per person and per office, are offered. This necessitated new work in areas such as operating systems, user interfaces, the internet, wireless, graphics, and a variety of other domains. At that time, Weiser called this work "ubiquitous computing." This is distinct from personal digital assistants (PDAs) and Dynabooks. UC that does not sit on a personal device of any kind, but is everywhere. Weiser believed that after 30 years, most interface design and computer design had followed the route of the 'dramatic' machine. UC's ultimate goal is to create a computer that is thrilling, magnificent, and fascinating. Weiser suggests the "invisible" path, which is a less-traveled path where the main aim is to make a computer so embedded, so comfortable, that the user stops worrying about it. Weiser believes that the second path will come to dominate in the next 20 years. But it won't be easy, PARC has been testing several versions of the future infrastructure. In the late 1990s, IBM's pervasive computing group popularized the phrase "pervasive computing" [8].

9.2.2 Characteristics of UC

The primary purpose of UC is to create linked smart products and make communication and data sharing easier and less intrusive [3]. The following are some of the key characteristics of UC:

■ Consideration of the human component and placement of the product in a human, rather than in a computational environment.
■ Using low-cost processors reduces memory and storage needs.
■ Obtains real-time data.
■ Computers are always connected and available.
■ Focus on many-to-many relationships rather than one-to-one, many-to-one or one-to-many, as well as the concept of technology, which is constantly present.
■ Includes local/global, social/personal, public/private, and invisible/visible characteristics, as well as knowledge development and information distribution.
■ Depends on internet convergence, wireless technologies, and modern electronics.
■ As digital gadgets become more wearable and networked, there will be limitations on and intrusion into user privacy.
■ The reliability factor of various equipment may be influenced as technology advances.

9.2.3 UC's Layers

HCI technologies such as IoT, Cloud Computing, and the others enable users to access wireless sensor networks. Such sensor networks gather data from a device's sensors before transferring it to a server. Three of the most common layers that it may encounter when transmitting are shown in Figure 9.1.

> *Layer 1: The Task Management layer* examines user tasks, context and index. It also manages the territory's complicated interdependencies.

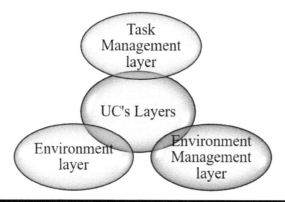

Figure 9.1   UC's layers.

Layer 2: The Environment Management layer monitors resources and their capabilities, as well as maps service needs and user-level status of specific capabilities.

Layer 3: The Environment layer keeps track of important resources and manages their dependability.

9.2.4 UC Types

UC can be categorized based on its portability, simplicity, technical capability, mobility, and accessibility. UC is widely divided into four categories, which are shown in Figure 9.2: (1) portable computing; (2) pervasive computing; (3) calm computing; and (4) wearable computing.

- *Portable computing*: Laptops and personal devices have made computing portable. Computers are now easily portable, allowing them to be carried wherever they are needed. Portable computers are used in a variety of industries and for a variety of purposes from the television industry to the military. Notebook and subnotebook computers, hand-held computers, palmtops, and PDAs are all examples of portable computers.
- *Pervasive computing*: Pervasive computing attempts to make our lives easier by providing tools that make it simple to handle data. Portable handheld personal assistant devices with high speed, wireless connectivity, lower power consumption rate, data storage in durable memory, small storage devices, color

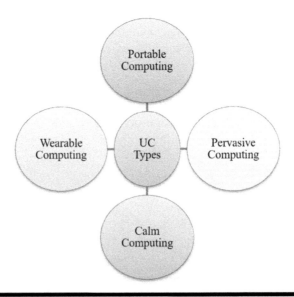

Figure 9.2   UC types.

display video, and voice processing technologies will be available through pervasive computing. Electronic toll systems on highways are examples of pervasive computing, as is monitoring software like Life360, which can track a user's position, speed, and how much battery life they have left.

■ *Calm computing*: Calm computing refers to a level of technical development in which the primary role of a user is not computing, but rather augmenting and bringing important information to the experience. Calm computing prioritizes people and tasks rather than computation and data. The next generation of linked devices is known as Calm Design. Its goal is to connect customers to their devices and other people in a way that allows them to live their lives without interruptions or pop-ups. A good example of calm technology is video conferencing. In contrast to telephone conferences, information can be gathered through gestures and facial expressions.

■ *Wearable computing*: The study of creating, designing, and managing computational and sensory systems is known as wearable computing. Wearable computing allows users to work, interact, and entertain while maintaining mobility and access to the device without using their hands or eyes. Wearables have the potential to improve interaction, learning, perception, and practical abilities. A wearable can filter user's calls, provide a reference, track user's health, remind user of a name and even assist users in catching the bus. Wearable technology refers to a group of electronic devices such as smart watches, rings, smart clothes, head-mounted displays, and medical devices that make life easier by connecting fitness, convenience, and lifestyle to the rest of the world [8].

9.3 UC Devices

Sensors are continually evolving, becoming more compact, lighter, more precise, durable, effective, reactive, and with increased communication capabilities. These key components, as well as the availability of new technologies, are supporting the expansion of the consumer electronics sensor industry, improving efficiency. For example, camera sensors are frequently used in smart systems to collect photos and videos of users interacting with the system. Digital color cameras can be used as sensing devices to interpret human hand positions, postures, and gestures, which can be converted into appropriate commands for controlling nearly any digital system. The capacity to recognize and track human movements and poses in the field is one of the most important features of camera-based HCI systems [9].

9.3.1 Smartwatches

Smartwatches are probably the best-known and most-used smart wearable in the workplace today. When a smartwatch is linked to a smartphone, the wearer can

Figure 9.3   Apple Watch.

Figure 9.4   Fitbit Luxe.

receive and send new messages directly from their watch, removing the need to hold and view the phone.

Apple's smart watches come in a variety of designs, with the features expanding with each new version (Figure 9.3). Safety features like fall detection, emergency SOS, and high and low heart rate notifications make the smartwatch useful for elderly family members, even if they don't own an iPhone. Calling, texting, and location sharing are simple ways to keep in touch with family through the smartwatch.

The Fitbit Luxe's lozenge configuration attracts the fashion-conscious, while still providing all of the health and fitness tracking, phone management, and even menstrual tracking that a regular wearable gadget provides (Figure 9.4). When compared with Apple smartwatches, the Fitbit is cheaper.

Figure 9.5   NFC Opn ring.

Figure 9.6   Oura Ring.

9.3.2 Smart Rings

Smart rings give consumers social feedback and can be used to interact with their surroundings in ways that other smartwatches and portable devices can't. When a text message, phone call, or other notification is received, certain smart rings provide notifications to alert the user.

The NFC Opnring allows the user to navigate apps, lock and unlock their door, and transfer data without being charged (Figure 9.5).

The Oura Ring delivers useful health information, such as sleep patterns and daily exercise (Figure 9.6). The Oura also monitors a variety of biometrics that many other wearable tech gadgets don't, and notifies the user if something is unusual.

9.3.3 Advanced Medical Wearables

Wearable medical tracking devices are designed to keep patients healthy and improve their quality of life. Wearables allow for continuous monitoring of physical activity quantity and quality, assisting patients in making well-informed changes in lifestyle.

Figure 9.7   Core body temperature sensor.

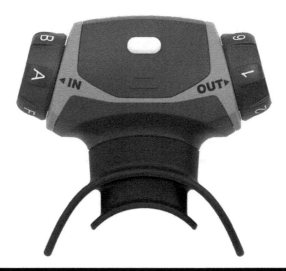

Figure 9.8   Airofit Breathing Trainer.

Heat training can help you improve your fitness or prepare for off-road, marathons, and Iron Man competitions. A core body temperature sensor can assist you in reaching your goal of successful heat training. Previously, extreme athletes could only track the metric with invasive measures like electronic pills (Figure 9.7).

Breathing training is an important therapy for asthmatics, athletes who want to improve their lung capacity, and individuals suffering from COVID-19. The Airofit Pro Breathing Trainer (Figure 9.8) and its companion software can help users to breathe faster and more efficiently by improving their lung function.

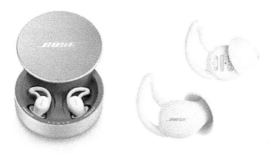

Figure 9.9   Bose Sleepbuds II.

9.3.4 Smart Earphones

Smart headphones, also known as smart earbuds or hearables, are already in audio devices that do more than just transmit audio. These headphones are often wireless and can connect to your phone, tablet, computer, or any other Bluetooth-enabled device.

Some research shows sleeplessness is one of the most common health problems. Sleep is taking a central role most of the time. The Bose Sleepbuds II Earphones have been professionally proven to improve sleep. They won't play radio or songs; instead, they'll play sounds from the Bose Sleep App (Figure 9.9).

9.3.5 Smart Clothing

Smart clothing can provide deeper insights than other forms of modern wearable technology since it makes contact with a larger area of the body, allowing for enhanced tracking for both medical care and lifestyle development. Samsung initiated substantial research in this field and has filed a number of interesting patents. If these patents become commercially accessible products, Samsung might produce smart shirts that diagnose respiratory problems and smart shoes that track running form, in the coming years. Consumers can already buy Siren Socks, smart socks that detect developing foot ulcers, Wearable X Nadi X smart pants (yoga pants that vibrate to improve form during yoga exercises), and Naviano smart swimsuits that provide alerts when the user should apply sunscreen (Figure 9.10).

9.4 UC's Applications

Nowadays, UC is a leading and unique technology that can be applied in a variety of research domains as well as other disciplines. This adaptable UC is used in a wide range of industries, particularly as sensors everywhere.

Figure 9.10   Smart clothes.

9.4.1 Healthcare Industry

UC is used in everyday hospital administration such as patient monitoring, well-being, and elderly care. Ubiquitous healthcare is a new field that uses ubiquitous technologies to provide a technology-oriented environment for healthcare professionals to provide efficient and effective care. Various technology solutions in diverse healthcare domains, such as chronic disease monitoring, gait analysis, mood and fall detection, neuropathic monitoring, physiological and vital sign monitoring, pulmonogical monitoring and so on, have been developed by the research community [10].

9.4.2 Accessibility

Gregg Vanderheiden used the term "ubiquitous accessibility" to describe a new technological approach to supporting accessibility. He suggests that, as computing evolves from personal workstations to UC, accessibility must be examined, based on this new computational paradigm. u-accessibility has arisen as a method of improving the quality of life of people with disabilities and the elderly.

9.4.3 Learning

Ubiquitous learning is defined as learning that is enhanced by the use of mobile and wireless communication technologies, as well as sensors and location/tracking mechanisms, all of which work together to integrate learners with their environments. To promote continuous, contextual, and meaningful learning, ubiquitous learning systems connect virtual and realistic things, people, and events. The technology dynamically helps the learning process by communicating with embedded computers in the surroundings while the learner is traveling with the mobile device. The objective of ubiquitous learning is to recognize which information can be delivered in different configurations and locations throughout the learners' everyday tasks and to link this data to the learners' educational process.

9.4.4 Logistics

Several studies have been conducted to assess the use of ubiquitous technologies to improve transportation logistics. MobiLab/Unisinos suggested two models in this context:

1. SWTrack: This enables businesses to track their vehicles and regulate the routes they travel. The model allows users to determine whether a vehicle is following a predetermined route or straying from it.
2. SafeTrack: This is a SWTrack extension that allows for the autonomous delivery management of loads with no user interaction.

RFID is used to manage the input/output cargo on the vehicle, identifying shipments and recoveries of loads made by mistake, as well as potential cargo theft, in real time.

9.4.5 Commerce

Context-aware commerce, also known as ubiquitous commerce or u-commerce, is the use of mobile devices and ubiquitous technology to promote business. Gershman identified three requirements for u-commerce success:

1. Consistently be linked with customers.
2. Always be conscious of clients' circumstances.
3. Always be responsible.

Some suggestions are focused on products trading, while others are concentrated on the market revenue. The MobiLab proposed the MUCS paradigm, which stands for "Universal Commerce Support." The UC technologies are used by MUCS to identify commercial opportunities for users as clients or suppliers. Furthermore, the model provides a generic method to encourage commerce in products and services across all domains.

9.4.6 Games

The Ubiquitous Games were born through the usage of UC technology in game production. In this type of game, players need to travel to certain locations in order to complete tasks. Furthermore, users interact with one another (multi-player) as well as with the environment around them (real objects). Wireless technologies enable users to walk while somehow sending and receiving information such as position, local users, and other types of context-aware data.

9.5 Advantages of UC

With the help of so many integrated sensors, UC is used in various fields to help people improve their day-to-day activities more efficiently. Some of the advantages are:

- Smart surroundings can be implemented with computing technologies that are primarily out of sight.
- Using smart networks, service costs can be reduced.
- Manufacturing productivity and scheduling are enhanced.
- In healthcare settings, response times are reduced.
- More accurate and convenient personal financial transactions.
- Individuals profit from the combination of sensors, networking technologies and data analytics to monitor and report on a variety of things, including purchasing preferences, production processes, and traffic patterns.
- Medical professionals can monitor patients' vital signs from a long distance, extending medical services remotely.
- The adoption of interactive media delivery technology, which allows students and teachers to communicate, can also help bring education to rural areas.

9.6 Disadvantages of UC

Even as technology advances, there are still some limitations. Some of the disadvantages of UC are discussed here:

- It could be challenging to implement while preserving acceptable privacy measures.
- Ubiquitous systems collect a lot of sensitive personal data and managing it includes dealing with legal, technical, and ethical issues.
- User consent could be a source of ethical concerns with UC.
- There should be clear security offered when it collects private information [10].

9.7 Augmented Reality (AR)

AR is a dynamically augmented version of the real physical environment created by the use of digital multimedia elements, music, or other sensory integration using technology. AR is defined as a system that combines real and virtual worlds, enabling real-time interaction and accurate 3D recognition of virtual and real things. The advance of AR technology is the best illustration of how HCI has innovated so much. AR is being used to improve natural environments and provide visually enhanced experiences. The information about the user's surrounding real-world environment becomes interactive and digitally modified with the use of modern AR technologies, such as computer vision, incorporating AR cameras into smartphone applications, and object identification. The real world is overlaid with data on the environment and its things. This data could be virtual. AR refers to any AR experience that augments the real world [11].

9.7.1 AR's History in a Nutshell

In 1968, Harvard professor and computer scientist Ivan Sutherland invented the first head-mounted display, named "The Sword of Damocles." Myron Kruger, a computer scientist and artist, established the 'Videoplace' laboratory at the University of Connecticut in 1974. The term "augmented reality" was invented by Tom Caudell, a Boeing researcher, in 1990. One of the first fully working AR systems was constructed in 1992 by Louis Rosenburg, a researcher at the USAF Armstrong Research Lab. Julie Martin, a writer and producer, pioneered AR in the entertainment sector with the theatre performance *Dancing in Cyberspace* in 1994. Sportsvision broadcast the first live NFL game using the virtual 1st & Ten graphic system, sometimes known as the yellow yard marker, in 1998. The technology overlays a yellow line on top of the footage so that viewers can see where the team just got a first down. NASA developed a hybrid synthetic vision system for the X-38 spacecraft in 1999. During their test flights, the system used AR technology to aid navigation. Hirokazu Kato created the ARToolKit, an open-source software library, in 2000. This package aids other developers in the creation of AR software. To overlay virtual images on top of the real world, the library uses video tracking. Sportvision upgraded the 1st & Ten graphics in 2003 to include the function on the new Skycam system, which provides spectators with an aerial shot of the field with graphics placed on top. For the first time in the media industry, *Esquire Magazine* employed AR to bring the pages to life in 2009. Volkswagen introduced the MARTA app (Mobile Augmented Reality Technical Assistance) in 2013, which provided technicians with step-by-step repair instructions in the service handbook. In 2014, Google Glass, a set of AR glasses that users might wear for immersive experiences, was presented. In 2016, Microsoft began delivering the HoloLens, a more advanced version of wearable AR technology than Google Glass [12].

9.7.2 Characteristics of AR

AR is a computer system that can merge real-world and computer-generated data and it has its own set of characteristics.

- It enables the actual and virtual worlds to be combined so that elements of the virtual world can interact with aspects of the actual world.
- Information is exhibited virtually, such as photos, audio, videos, graphics and GPS data, and is strongly linked to the information that we see with our actual eyes.
- It should be real-time interactive.
- It directly interacts with the environment's physical capabilities.

9.7.3 AR Types

Many immersive AR app development companies are always looking to improve the consumer experience by integrating rich and innovative ways of expressing digital content. This is made feasible through the use of several types of AR.

9.7.3.1 Marker-Based AR

Marker-based AR uses AR applications to identify physical images, 'Markers' and 3D models are shown in Figure 9.11. When you open the AR app, it uses your mobile device's back camera to track such markings. This form of AR technology is also known as image recognition or recognition-based AR. AR displays overlay content in the form of video, image, 3D model, or animation clips,. Users can connect with this content through an app. This is a popular sort of AR that allows users to observe an object or image in better detail and from various angles. Furthermore, when the user turns the marker, the 3D imaginary rotates at the same time [13].

9.7.3.2 Markerless AR

Markerless AR works by scanning the environment without the use of a marker (Figure 9.12). Users can place the virtual object or information anywhere they want without having to move anything in the background. The markerless AR is a type of virtual reality that collects positional information using the device's location, digital compass, camera, and accelerometer. In addition, for AR products to not always float in the air, mobile apps with such features frequently ask the user for a flat surface or floor to place them on. The AR app detects that flat surface and places items on top of it [13].

Figure 9.11 Marker-based AR.

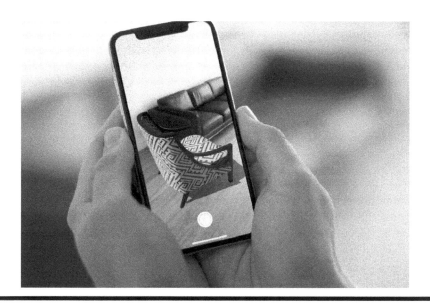

Figure 9.12 Markerless AR.

Figure 9.13 Location-based AR.

9.7.3.3 Location-Based AR

This is one of the most frequently used types of AR. The location is mostly determined by GPS, Digital Compass, smartphone camera, and other technologies. It does not require special markers to identify the location where the virtual object is placed, unlike marker-based AR (Figure 9.13). A specific location is assigned to the virtual products. The objects are displayed on the screen when the user enters a predetermined area. Because it can forecast the user's focus to connect real-time data with the current spot, location-based AR does not require a hint from the object to deploy. It also enables app developers to display imaginative, interesting, and valuable digital content to physical points of interest [13].

9.7.3.4 Superimposition AR

Superimposition AR creates a different view of an object that can be used to temporarily replace the original view (Figure 9.14). This means that an augmented view replaces the full view of an object or a portion of it with this technology. Object recognition is key in this type of AR. Superimposition AR displays numerous perspectives of a target object with the option of emphasizing additional relevant information [13].

9.7.3.5 Projection-Based AR

Projection-based AR is a video projection technique that allows users to extend or distribute digital data by projecting images onto the surface of 3D objects or into their actual space (Figure 9.15). The user does not operate this type of technology.

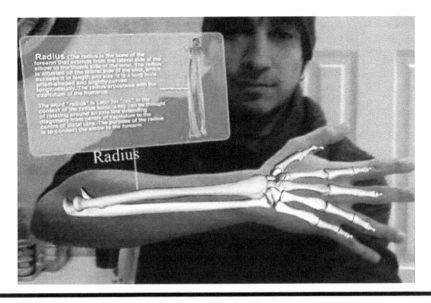

Figure 9.14   Superimposition AR.

Figure 9.15   Projection-based AR.

This is one of the most useful types of AR since it allows users to freely navigate around the world within the limitations of the projector and camera. With this technology, it is simple to create graphical representations with high-definition photos or films that are impossible to see with traditional lighting techniques. With the passage of time, it can also modify the visual appearance of the thing. This is AR software that uses artificial light to create illusions of depth, position, and orientation of objects on real flat surfaces under proper supervision to simplify difficult manual activities in organizations. The best part is that because the instructions are presented on a specific work location, it eliminates the need for computers and screens [13].

9.8 AR Devices

AR is often recognized as one of the most promising technologies currently available. After the smartphone revolution, everyone carries a mobile device and all of these mobile devices generally include a processor, GPS, a display, a camera, microphone, and some other hardware essential for AR. The terms AR and VR are frequently misunderstood. When it comes to AR, while most smartphones and tablets can handle it, VR requires specialized hardware.

9.8.1 Microsoft HoloLens

Microsoft HoloLens is an AR headset with transparent lenses. AR content is added to the environment using a set of sensors and highly developed lenses. The HoloLens also includes multiple microphones, an HD camera, a light sensor, and the "Holographic Processing Unit," which allows users to interact with 3D holograms that are presented as part of the real environment. The ability to interact with the virtual world, in particular, provides customers with a wide range of possibilities while using this technology. Automobile manufacturers, for example, could use the HoloLens to make real-time changes to new models (Figure 9.16). Other examples include watching Netflix on your room's wall or having a virtual pet wander around your room.

Users can use their eyes to choose hologram apps and interact with the VR world. The cursor will follow the movement of the head and body gestures can be used to pick holograms or products and activate programs. To navigate and operate apps, use voice recognition.

9.8.2 MagicLeap One

The "Lightpack," a little computer that can be attached to your belt or pocket, powers this futuristic-looking AR headset. MagicLeap's ultimate goal is to combine your virtual and actual lives. MagicLeap, for example, can project a whale into a user's classroom (Figure 9.17). MagicLeap does more than just insert image content in the real world. The computer-generated stuff, like the HoloLens, responds to its environment and allows the user to communicate with it. The way the user interacts with virtual content differs. MagicLeap employs a controller with a large button and a touchpad instead of gestures and eye movement.

Figure 9.16 Designing a car by using HoloLens.

Figure 9.17 Whale in a MagicLeap.

Figure 9.18 Drone scene with Epson Moverio.

9.8.3 Epson Moverio

This AR headset has been specifically developed to operate with binoculars and transparent glasses. This may not be the best choice if the user is seeking a fashionable and elegant headset. The Moverio glasses, on the other hand, are extremely adaptable and can be adjusted to fit any size or user requirement. Epson Moverio's Si-Oiled technology produces a sharp, bright, and high-quality image for customers. These smart glasses operate on Android 5.1 and are powered by an Intel Axom 5 CPU, making it easier for developers to create AR apps. Flying drones with the Epson Moverio headset's AR app is a fun way to use it (Figure 9.18). The user can have a hands-free flying experience while also seeing the drone through the glasses.

9.8.4 Google Glass Enterprise Edition

Google's hands-free and wearable computers were created to assist users in offering excellent productivity experiences. Maintenance technicians are able to concentrate on the task at hand while still having access to a manual. Mistakes and distractions can be considerably decreased by minimizing this constant focus shifting. Google Glass Enterprise is a follow-up to Google Glass and Google Glass Explorer Edition. A lighter device, with longer battery life, a faster CPU, and a camera upgrade from 5 to 8 MP are the main features of the Google Glass Enterprise Edition (Figure 9.19).

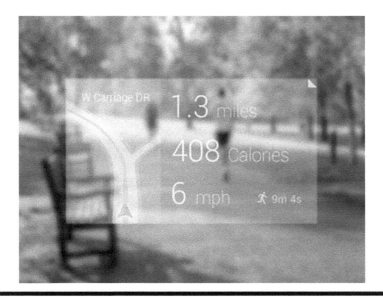

Figure 9.19   Activity information provided by Google Glass.

These AR headsets and smart glasses have the potential to take over the AR world [14]. However, smartphones, mobile devices, and AR smart glasses will very certainly interact in the future.

9.9 AR Applications

AR and other VR technologies are becoming increasingly popular. Here are modern-day AR applications that users should be aware of.

9.9.1 In the Military

Microsoft is working with the US Army to develop the Integrated Visual Augmentation System (IVAS) to increase soldiers' situational awareness, communications, battlefield navigation, and overall operational effectiveness. IVAS has a heads-up display (HUD), thermal imaging, interactive maps, and an overhead compass, all of which are powered by Microsoft's HoloLens technology (Figure 9.20). Soldiers can use it to track and share enemy positions across the battlefield. Additionally, it can identify whether a target is dangerous, moderate, or helpful. Multiple front-facing head-mounted cameras can see in the dark (night vision), see through smoke, and even look around corners [15].

Figure 9.20   AR in the military.

9.9.2 3D Animals

Another exciting application of AR technology is 3D animals. Google 3D Animals allows users to search for and watch life-sized animals in 3D, as well as interact with them via AR in their local area. Google image-search looks for 3D creatures like tigers, zebras, koalas, or sharks using an ARCore-equipped device and supported browsers like Chrome, Opera, and Safari. Use "Tap on View in 3D," then view in your space on the results page. To see the life-sized tiger in 3D, simply point the smartphone to the ground (Figure 9.21). The user can rotate the animal and hear the animal bark, bray, and snort, much like it does in the wild, using AR. This was launched at Google I/O 2019 and includes a growing number of animals.

9.9.3 Fashion

See My Fit, a technology developed by Asos, uses AR to visually fit garments on models. This allows the user to try up to 500 things per week on six different models to see how they fit. The user can visualize the size, fit, and cut of each item before making a purchase. Due to COVID-19, the shop is now unable to work with models in its studios. In addition to See My Fit, Asos is integrating flat shot photographs and product shots from models' homes to continue to serve its customers. The Virtual Catwalk, an AR-enabled tool that allows users to visually envision and experience up to 100 ASOS Design products, was introduced earlier this year by the top fashion

Figure 9.21   Google's 3D Tiger AR image.

store (Figure 9.22). Zara, Ikea, Shopify, Kendra Scott, and a slew of other retailers are among the fashion companies aiming to use this technology in future.

9.9.4 Gaming

In game creation, game creators are experimental with more immersive and experiencing elements. For example, Pokémon Go, which was introduced in 2016, is one of the most prominent examples of AR. It allows users to experience AR by bringing game characters into their physical surroundings (Figure 9.23).

9.9.5 Coloring Books

Traditional coloring books allow users to sketch and express themselves artistically in two dimensions on paper. However, in today's extended reality environment, this is static and fairly restricting. Use AR to explore the amazing world of 3D drawing. Users can sketch their favorite characters in 2D on paper and have them reproduced in 3D in real time with Disney's AR coloring book app (Figure 9.24). The program matches and copies elements of the drawing in 3D using texture generation. You can move your characters as you turn the pages using deformable surface tracking. Coloring books by Disney are accessible for both children and adults, with over 100 illustrations to encourage creativity and motivate you to draw. Disney received the Edison Award in 2016 for this unique application.

Figure 9.22 AR hopping.

Figure 9.23 Pokémon Go AR game.

Figure 9.24   Disney's AR Coloring Book.

9.9.6 Obstetrics

Road to Birth is an AR+VR project being created by the University of Newcastle's Innovation Team to provide visual insights into the stages of childbirth and their effects on pregnant women. This is a world-first invention that is designed to improve healthcare for both healthcare professionals and pregnant women. Samsung GearVR, HTC Vive (VR) and HoloLens are all used (AR). Using AR and VR, this mixed-reality simulation (Figure 9.25) will allow you to learn about essential anatomy changes, delivery approaches, and real-world situations in 3D. This will help students learn faster and provide pregnant women with a better understanding of the biological changes they will experience throughout pregnancy.

9.9.7 Architecture

In the architecture world and construction industries, AR is used for space planning and design visualization. Turning a simple sketch design on paper into a full 3D model with specialist software is a great advantage of AR.AR can also be used in construction to take a virtual tour of the finished 3D model during design analysis (Figure 9.26). This visual inspection can support engineers and contractors in determining the height of the ceiling, as well as identifying any construction difficulties and assisting with component prefabrication. The University of Canterbury in New Zealand launched CityViewAR in 2011, following the Christchurch earthquake, to

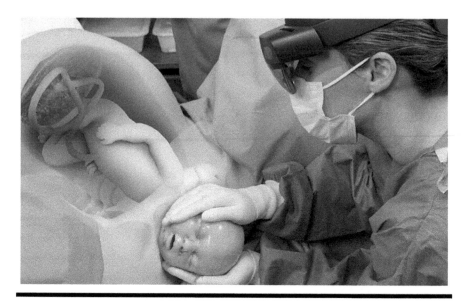

Figure 9.25   Childbirth healthcare training by AR.

Figure 9.26   AR in architecture.

Figure 9.27   AR in the sports industry.

allow city planners and engineers to understand and re-imagine structures as they were before the earthquake.

9.9.8 Sports

Consider 3D team set-plays, pre-game practice with computer-generated opponents, virtual training venues, custom-built training sets for individual athletes, and more (Figure 9.27). These technologies are already in use by teams in the National Football League (NFL), the Atlantic Coast Conference (ACC), and several soccer teams. One of the companies driving this trend is EON Sports. Virtual training and 24/7 personal AR coaches promise fewer injuries in sports, as well as greater experiences for athletes, officials, spectators, and other stakeholders.

9.10 Advantages of AR

The possibilities are enormous as AR technology becomes more accessible and affordable.AR is a rapidly expanding technology with a wide range of applications. The following are some of the advantages:

- Reduces the gap between the actual and virtual worlds.
- Enhances perceptions of the real world and interactions with it.
- Patients' lives have been safer as a result of AR use in the medical field. It aids in the accurate diagnosis and early detection of disorders.

- With the apps, anyone can use them, as they are easy to manipulate.
- Saves money by putting crucial circumstances to the test before adopting them in real life.
- Can be implemented in the current world once it has been verified.
- Military personnel can use AR in battlefield simulations before the actual conflict without putting their lives at risk. This will also support them in making important decisions during the actual fighting.
- Can be used as part of training programmes because it makes things memorable and interesting [16].

9.11 Disadvantages of AR

Every new piece of technology that is emerging also has a negative aspect to it. The following are the disadvantages of AR:

- Costly to build and sustain AR technology-based projects.
- Lack of privacy is a major issue.
- People are missing out on critical moments in AR.
- Low performance is a problem that must be addressed during the testing process.
- To efficiently use AR-compatible devices, the user needs some fundamental knowledge [16].

9.12 Conclusion

The field of HCI covers a number of disciplines, including computer science, cognitive science, design, mass communication, and others. The two major HCI developments discussed in this chapter seem to be UC and AR. This chapter provides a comprehensive overview of UC, detailing it from its early years to the present. The UC timeline, characteristics of UC, layers of UC, types of UC, and various modern technologies used by people, such as smartphones, smart watches, smart rings, and so forth, have been properly covered. This chapter described the applications of UC in the fields of healthcare, education, logistics, business, and the gaming industry. UC's benefits and drawbacks were listed. Similar to UC, AR's history, characteristics, types, AR devices, AR applications, and advantages and disadvantages were all explained. AR is a revolution for the HCI economy. The level of progress and continued development of the technology are visible to the public.

Acknowledgments

We thank God Almighty for His abundant grace and mercy who made this possible.

References

1. MRCET (n.d.) "Digital Notes on Human Computer Interaction." Available at: https://mrcet.com/pdf/Lab%20Manuals/IT/R15A0562%20HCI.pdf (accessed May 31, 2022).
2. Fakhreddine Karray, et al. (2008) "Cooperative Multi Target Tracking Using Multi Sensor Network." *International Journal on Smart Sensing and Intelligent Systems*, 1(3).
3. Wikipedia (n.d.) "Human–Computer Interaction." Available at: https://en.wikipedia.org/wiki/Human%E2%80%93computer_interaction. (accessed August 1, 2021).
4. The Interaction Design Foundation (2015) "The 5 Main Characteristics of Interaction Design | Interaction Design Foundation (IxDF)." 1 August. Available at: www.interaction-design.org/literature/article/the-5-main-characteristics-of-interaction-design.
5. Educative (n.d.) "Introduction to Human-Computer Interaction & Design Principles." Interactive Courses for Software Developers. Available at: www.educative.io, www.educative.io/blog/intro-human-computer-interaction. (accessed May 31, 2022).
6. "Human Computer Interface–Quick Guide." Available at: www.tutorialspoint.com, www.tutorialspoint.com/human_computer_interface/quick_guide.htm (accessed May 31, 2022).
7. Utsav Mishra, and AnalyticsSteps (2021) "Human-Computer Interaction (HCI): Importance and Applications." June 25. Available at: www.analyticssteps.com/blogs/human-computer-interactionhci-importance-and-applications.
8. IoT Agenda (n.d.) "What Is Ubiquitous Computing (Pervasive Computing)?",1 Oct. 2019,Available at: www.techtarget.com/iotagenda/definition/pervasive-computing-ubiquitous-computing.
9. Wikipedia (n.d.) "Portal: Human–Computer Interaction." Available at: https://en.wikipedia.org/wiki/Portal:Human%E2%80%93computer_interaction (accessed 31 May 2022).
10. IoT Agenda (2019) "What Is Ubiquitous Computing (Pervasive Computing)?"1 Oct. Available at: www.techtarget.com/iotagenda/definition/pervasive-computing-ubiquitous-computing.
11. The Franklin Institute (2017) "What Is Augmented Reality?" Sept. 21. Available at: www.fi.edu/what-is-augmented-reality.
12. www.G2.Com/Articles/History-of-Augmented-Reality (accessed May 31, 2022).
13. Ayushi, Doegar, et al. (2021) "What Are the Different Types of Augmented Reality? Latest Technology Blogs &Updates by Top Research Agency." Oct. 27. Available at: https://mobcoder.com/blog/types-of-augmented-reality/.
14. Onirix (2020) "The Best Augmented Reality Hardware in 2019–Onirix." May 15. Available at: www.onirix.com/learn-about-ar/the-best-augmented-reality-hardware-in-2019/.
15. Ritesh Pathak, and AnalyticsSteps (2020) "7 Top Trends in Augmented Reality | Use Cases of AR." October 20. Available at: www.analyticssteps.com/blogs/5-trends-augmented-reality.
16. Myayanblog (n.d.) "Advantages and Disadvantages of AR–Augmented Reality." Available at: www.myayan.com/advantages-and-disadvantages-of-ar-augmented-reality (accessed May 31, 2022).

Chapter 10

A Machine Learning-Based Driving Assistance System for Lane and Drowsiness Monitoring

Sanjay A. Gobi Ramasamy

10.1 Introduction

Lane detection is a critical component when driving heavy vehicles. This is done by using computer vision methods with the help of the OpenCV library [1]. Lane detection is performed by looking for white markings on both sides of the lane and patterns with color. The number of people driving along the nation's highways without proper sleep is uncountable, causing accidents. Using Python, OpenCV, and Keras, the driver will be alerted to drowsiness and avoid accidents [2]. The concept of driving in the right lane refers to avoiding the risk of moving into another lane without the knowledge of the driver, which leads to accidents [2]. Lane departure signals and lane-keeping signals are among the warning signals. Sleep deprivation is a problem for truckers who travel long distances. As such, driving while sleepy becomes extremely dangerous as the driver can wander into the wrong lane, so we recommend a portable system to monitor a driver's actions that can sit on the windshield or any surface.

Humans have developed technology to simplify and protect their lives, whether they are performing mundane tasks, such as going to work [2] or more complex ones, such as traveling by air. Technology has impacted our lives profoundly. At the pace we travel today, even our grandparents wouldn't have believed we could do it.

DOI: 10.1201/9781003424550-10

193

Almost everyone on this planet now uses some type of transport every day. Others use public transportation because they cannot afford a car. No matter a person's social standing, there are a few rules for drivers. For instance, remaining alert and active while driving is a must [3]. In developing countries such as India, fatigue, combined with bad infrastructure, leads to disaster. There are no clear key indicators or tests available for measuring or observing fatigue, unlike alcohol and drugs, which have obvious key indicators and tests [4]. It is likely that raising awareness of fatigue-related accidents and encouraging drivers to report fatigue at the earliest opportunity can solve this problem. Many long-distance truckers have to drive for many hours, which is challenging and costly. But the higher wages associated with the job encourage many people to take on that job. Transport vehicles must be driven at night as in the day. When drivers are motivated by money, they make unwise decisions, such as driving all night despite being fatigued [3]. Driving while fatigued causes many accidents since the drivers don't realize the risk involved.

To solve this problem, however, it is very difficult and costly to implement these restrictions, and, even then, it is not sufficient to solve this problem. Some nations have imposed restrictions on the number of hours a driver can drive at a stretch [3]. Aesthetic and technical enthusiasts who wish to develop the project are the intended audience for this chapter. In many vehicles, there are a variety of products to measure driver fatigue. Unlike the traditional drowsiness detection system, our driver drowsiness detection system offers better results and additional benefits. A user will receive an alert if the measure of drowsiness reaches a certain saturation level.

Despite fatigue being a health and safety problem, no nation in the world has yet managed to tackle it comprehensively [5]. Alcohol and drugs, on the other hand, have easy-to-measure key indicators and tests, while fatigue is difficult to measure or observe. It is impossible to discern fatigue without the some sort of indicator, as the drivers will not state their fatigue level, due to the high-profit margin of driving long hours.

10.2 Literature Review

10.2.1 System Review

This survey examined various sites and applications and looked at the fundamental data to figure out what the general public's needs are. By analyzing these data, we came up with an audit that gave us a fresh perspective and helped us to devise different strategies. A similar application is needed and some reasonable progress is being made in this area.

10.2.2 Lane Detection Techniques

An extended convolutional neural network (CNN) was created by developing a deep multitasking network that takes into account two neural networks and uses deep learning to detect lanes. CNN and recurrent neural networks (RNN) are both effective in detecting lanes. As part of this process, a camera attached to the vehicle takes a picture of the road. Next, the image is converted to grayscale to reduce the processing time. Following that, if the image suffers from interference, one can activate filters to remove these interferences to ensure accurate edge detection. In addition to bilateral and Gaussian filters, trilateral filters can also be used. Using a canny filter and an edge detector, an edged image can be obtained using thresholding generated by a machine to obtain the edges. These edges can be detected by line detectors. Segments of the lane boundary will be generated on the left and the right. As a result, the RGB color codes are used to determine yellow and white lanes.

10.2.3 Robust Lane Detection in Low Light

To remove unnecessary objects from an image, an original image needs to have a Region of Interest (ROI). That is, it should remove things like street lights, signs, the moon, etc. By cropping, the lane detection system can also be made more accurate and faster. As the image size decreases, the amount of velocity enhancement is increased. Accuracy can be improved by nutrients objects which have the same characteristics as lanes but are located outside of the ROI. As implemented here, the ROI is set manually; however, camera calibration parameters could be used to automatically calculate a suitable ROI. This lane detection system only requires pre-processing of color images to create grayscale images; therefore, this step can be skipped for monochrome images. For the accessible image, the Bayer format is assumed. A pixel's color should be extracted in the first step. By contrast, shoulder lanes are generally more visible in contrast to a traffic lane because they are typically presented in the image as long, straight lines.

10.2.4 Requirement

Figure 10.1 shows the tools required to carry out the process.

10.2.5 Use Case Diagram

Figure 10.2 shows the use case diagram.

The incremental model is used to create the framework. In this manner, the framework's center model is first established, and then, after every testing stage, it is augmented accordingly. Project skeletons were optimized to expand capability.

10.2.6 Process Flow

Figure 10.3 shows the process flow.

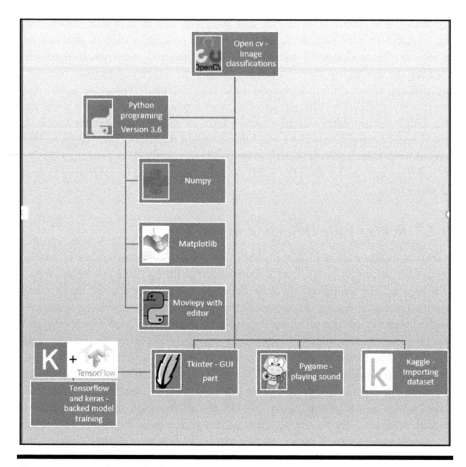

Figure 10.1 Tools required.

10.3 Framework for Performance Analysis

Detecting a road from a single image involves passing the image through a computer vision algorithm, which then outputs the marked spots on the image. During automatic driving, an image is captured by the car's driving camera, which is then used in the architecture (Figure 10.4). The computer is supposed to automatically recognize the road, which should include all the details of the model [5]. A road model consists mainly of road edges and road lines. Edge detection will use a canny edge detection algorithm and road line detection will use a Hough transform space algorithm.

10.3.1 Working on Drowsiness Detections

Figure 10.5 presents the system of how the model works and Figure 10.6 shows the model training stages.

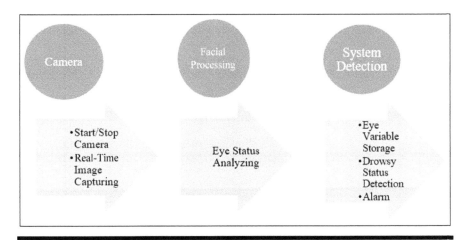

Figure 10.2   Use case diagram.

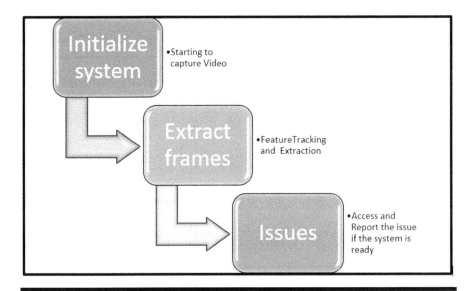

Figure 10.3   Process flow.

10.4 Image Capturing

Webcams are used to capture input images. An infinite loop records every frame from the webcam while it is accessed. The cv2.VideoCapture(0) method of the OpenCV library is used to capture video and set the capture object [3]. Each image is stored in a video frame variable by reading () in the frame.

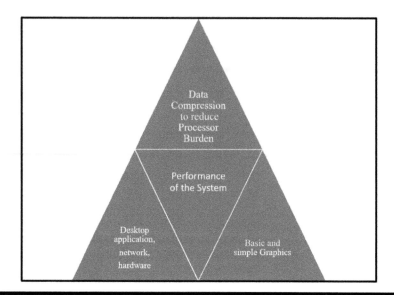

Figure 10.4   The performance analysis framework.

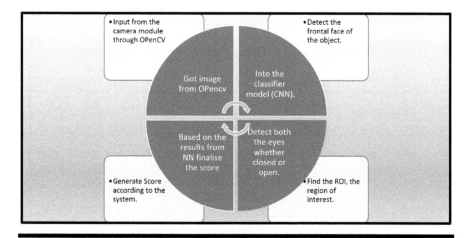

Figure 10.5   Workflow of the overall system.

Grayscale images are required for OpenCV's object detection algorithm to detect a face in the image. Detection of faces is not dependent upon color information [3]. This is done by implementing a Harr cascade classifier.

10.4.1 Edge Detection

Boundaries are defined by edges, which makes image processing a fundamental problem. An edge is a sharp contrast between different pixels, a sudden change in

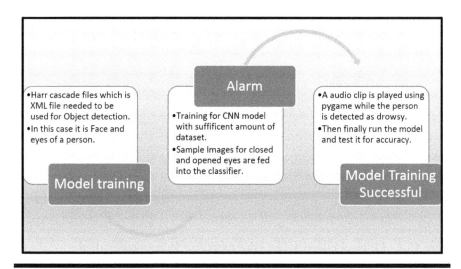

Figure 10.6   Neural network model training.

intensity [5]. When edges are detected in an image, data is significantly reduced and useless information is culled, while the image's significant structural properties are preserved. One example is an optimal edge detector, also called canny. Ideally, the first criterion should have a low error rate and make sure that useful information is preserved while unwanted data is filtered out. As much variation as possible should be kept between the original image and the processed image as the second criterion [5]. The third criterion involves removing multiple responses to edges. The edge detector uses these criteria to smooth out the image first. Next, it highlights regions with high spatial derivatives by calculating the image gradient.

Based on these regions, the algorithm then suppresses any pixels that are not at the maximum using non-maximum suppression. To remove streaking along the edges and then the gradient array, hysteresis is now applied. Figure 10.7 shows the original image edge detection techniques applied. Figures 10.8 and 10.9 show the Sobel enhanced technique.

10.4.2 Feature Extraction Frontal Face

In this function, the classifier can recognize facial characteristics based on the Cascade Classifier's function. This is followed by the function detect multiscale with grayscale image which detects faces on multiple scales [6]. Detection arrays are returned with coordinates, as well as height, which indicates the width of a box that defines the boundary of an object. This makes it easier to iterate over each face and draw boundary boxes. In the first step, both eyes are assigned to a

Figure 10.7   Original image before edge detection techniques applied.

Figure 10.8   Sobel enhanced edge in X, Y direction.

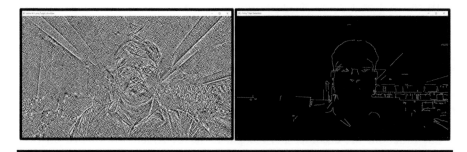

Figure 10.9   Canny edge detection and Sobel enhanced edge in X, Y direction through Sobel function.

cascade classifier and then each cascade classifier is used to detect both eyes [6]. The full image in this case only needs to be analyzed for eye information. The boundary box of the eye must first be extracted so that the eye image can then be grabbed. Only the left eye's image data is contained [6] in the function detects. In this way, we will be able to determine whether an eye is open or closed by using our CNN classifier. This process is repeated for determining whether the right eye is open.

- Convolutional neural networks (CNN) were used to build the model using Keras.
- An image classification algorithm that uses convolutional neural networks is exceptionally effective for recognizing images.
- This model has three layers: one input layer, one output layer, and one hidden layer.
- To perform a convolution on these layers, a filter multiplies the layer with a filter of 2D matrices.
- The final layer has two nodes and is fully connected.
- All layers are activated using Relu activation functions, except for the output layer where Softmax is used (Figure 10.10).

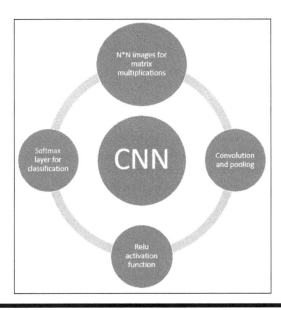

Figure 10.10   The model architecture.

Figure 10.11   Screenshot of eyes closed and warning score.

10.4.3 Grayscale Conversion

A few operations are required to fix the corrected dimension before the image can be fed into the model. Using the cvtColor function with color from BGR to grayscale attribute to convert a color image into grayscale. The model was trained on 24*24-pixel images [7], so the image is resized to 24*24 pixels. If you want better convergence, normalize the data by multiplying the right eye by 255 (all values will be within the range of 0–1). To feed the classifier with more dimensions, expand the data [8]. Use the model load_model function to load the model. The model predict_ classes class will be used to start the process of classifying each eye with the model [8]. Having the predicted class value of 1 indicates that the eyes are open, and having the predicted class value of 0 indicates that the eyes are closed.

10.4.4 Score Calculation

As a result of the score, we can calculate how long the individual has had their eyes closed. As long as both eyes are closed, the system will constantly increase the score, while as long as the eyes are open, the score will decrease (Figure 10.11). After the classification has been made on the screen, Cv2.putText will display the status of the person in real time. If the score is above 15, for example, then the person's eyes must have been closed for a long time; the system will subsequently beep the alarm using the sound_play function.

10.5 Proposed Model for Lane Detection System

Urban traffic has become more and more complex when it comes to traffic safety. Most accidents on the roads are caused by exiting lanes without following the

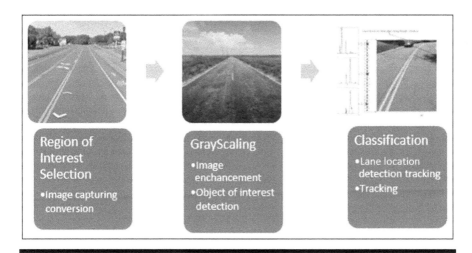

Figure 10.12   Model for lane detection system.

proper rules. The majority of these crashes occur because the driver is lethargic and interrupted. Regardless of whether the driver or pedestrian is distracted, lane discipline is crucial. The software has the goal of identifying lane markings [3]. This software will improve traffic conditions.

The changes in lane boundaries have caused the above-stated problems. In this chapter, an algorithm is proposed for the detection of lane markings on the road by using computer vision technology to analyze the video of the road [5] (Figure 10.12). This technology aims to reduce the frequency of accidents by detecting lane markings on the road.

To prevent accidents caused by reckless driving on the road, the system can be installed in cars and taxis. The system will ensure the safety of children in school buses. The set-up can also monitor the performance of drivers, which in turn allows the Road Transportation Offices to report and check negligence and lack of attention by drivers on the road.

10.5.1 Comparing the Accuracy of the Lane Detection System

The arrangements bounded by the type of traffic environment and the roads noted above must be taken into account when computing a practical collection of driving conditions that a driver will encounter while driving in a darker environment. Several conclusions can be drawn. Video acquired over five hours is used to evaluate the performance of the proposed lane detection system.

Frame-wise access permits flexible conclusions to be drawn when analyzing videos with varying frame rates. Video with varying frame rates can be evaluated better using this measurement rather than other methods.

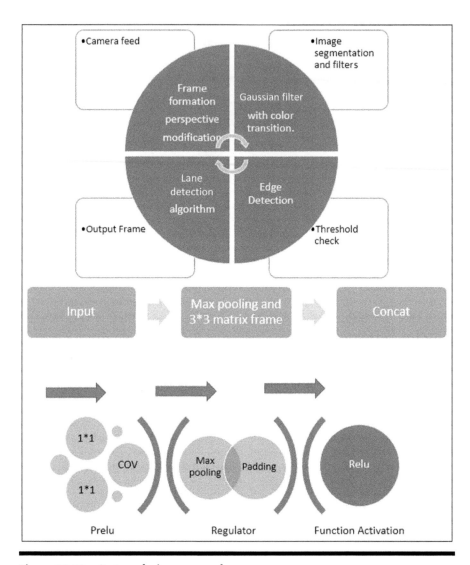

Figure 10.13   System design approach.

10.5.2 Implementation

10.5.2.1 System Design Approach

Figure 10.13 shows the system design approach.

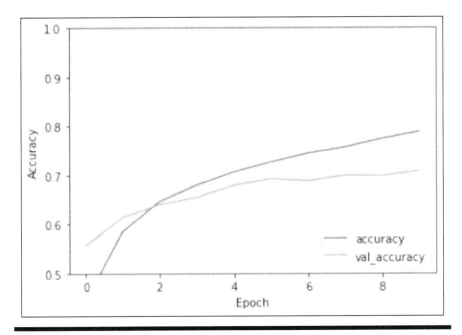

Figure 10.14   Model accuracy graph.

10.6 Computational Results

Image segmentation is done pixel-by-pixel using semantic segmentation. Datasets used for training were downloaded from the Internet. Segmentation was performed using 30 classes. Prelu is used for simple convolutions diluted with activation functions. The main goal is finding the kernel.

10.6.1 Accuracy Achieved

Here the model is trained as a steady line in the graph which indicates that the model trained is efficient which both the value-added points (Figure 10.14). An accuracy check of the system can be carried out by measuring the system efficiency per minute (Table 10.1). As the accurate column indicates, Table 10.2 shows accurate detections per minute, while the Approximate and Missing columns in Figure 10.15 show approximately positives and undetected lanes, respectively.

10.6.2 Assumptions

Imagine a sunny day with adequate lighting on the roads. Lane detection systems are assumed to work up to speed and not limit the memory. The yellow lane markings mark the left side of the road whereas the white lane markings mark the right.

Table 10.1   Model Accuracy

Epoch range	Loss	Accuracy	Val_loss	Val_accuracy
1/10,1563/1563	1.4971	0.4553	1.2659	0.5492
2/10,1563/1563	1.1424	0.5966	1.1025	0.6098
3/10,1563/1563	0.9885	0.6539	0.9557	0.6629
4/10,1563/1563	0.8932	0.6878	0.8924	0.6935
5/10,1563/1563	0.8222	0.7130	0.8679	0.7025
6/10,1563/1563	0.7663	0.7323	0.9336	0.6819
7/10,1563/1563	0.7224	0.7466	0.8546	0.7086
8/10,1563/1563	0.6726	0.7611	0.8777	0.7068
9/10,1563/1563	0.6372	0.7760	0.8410	0.7179
10/10,1563/1563	0.6024	0.7875	0.8475	0.7192

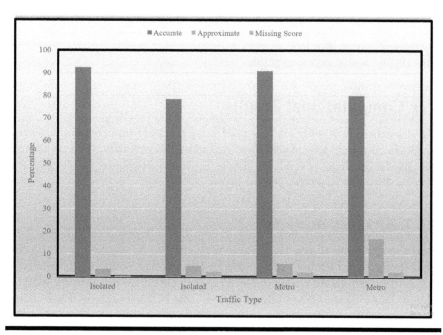

Figure 10.15   Graph for accuracy.

Table 10.2   Accuracy Table

Accuracy table (Average value per minute)				
Highway type	*Traffic type*	*Accurate (%)*	*Approximate (%)*	*Missing score (%)*
Isolated	Light	92.3555	3.45	0.34
Isolated	Moderate	78.4352	4.097	2.34
Metro	Light	91.345	5.897	1.987
Metro	Heavy	80.0054	16.897	2.23

Highway authority standards should apply to all roadways built. It is assumed that the lane detection system is fast enough and that there is no limit to memory.

10.7 Conclusion

In addition to its limitations, OpenCV is inefficient and, without clear markings, produces inaccurate results. The classical OpenCV method cannot accurately detect roads without clear markings. This method is also incompatible with all environments. The model can be tuned incrementally by adding additional parameters, such as blink rate, car condition, and yawning rates. A variety of applications have been developed to address these issues, including traffic control, traffic monitoring, and traffic flow.

The modular implementation allows us to update algorithms easily and continue working on changes to models in the future. Using a pickle file of the model, the needed areas can be inserted and then easily applied to the product. Consequently, it would not be necessary to compile the entire code each time. By introducing the concept of detecting dark roads during the night, we can add value to the project.

When it is daylight, the process identifies and selects colors very well. The accuracy can be significantly improved by using all of these parameters. Netflix and other streaming services can use the same model and techniques to detect when users are asleep and stop their videos accordingly. Anti-sleep applications could also be developed using this method. Poorly functioning autonomous driving systems have been implicated in several high-profile accidents, but other accidents have shown the benefits of this technology. China has reported a fatality after a Tesla driver hit a cleaning vehicle with his vehicle in January 2016. At the time of the crash, Tesla's autonomous features were reportedly engaged. In this first death involving an ADS feature, the Tesla driver was not paying enough attention to the road, according to the police.

In terms of safety, there are still difficulties evaluating self-driving cars using publicly available data. Driverless cars are usually tested in cities and states where the weather is relatively dry and the road system is simple, which facilitates autonomous driving. California is also the only state in America that requires detailed accident reports from companies testing driverless cars. For this reason, it is also the only state that requires detailed accident reports for any accident involving autonomous vehicles on public roads.

References

1. J. Long, E. Shelhamer, and T. Darrell (2015) "Lane Detection Techniques – A Review." In *Proceedings of the IEEE Conference on Computer Vision and Pattern Recognition*, pp. 3431–3440.
2. S. Zheng, S. Jayasumana, B. Romera-Paredes, V. Vineet, Z. Su, D. Du, C. Huang, and P. H. Torr (2015) "A Layered Approach to Robust Lane Detection at Night." Conference paper, pp. 1529–1537.
3. P. Shopa, N. Sumetha and P. S. K. Pathra (2014) "Traffic Sign Detection and Recognition Using OpenCV." Paper presented at International Conference on Information Communication and Embedded Systems (ICICES2014).
4. V. Badrinarayanan, A. Handa, and R. Cipolla (2015) "Segnet: A deep Convolutional Encoder-Decoder Architecture for Robust Semantic Pixelwise Labelling." *arXiv* preprint arXiv:1505.07293,
5. W. Han, Y. Yang, G.-B. Huang, O. Sourina, F. Klanner, and C. Denk (2015) "Driver Drowsiness Detection Based on Novel Eye Openness Recognition Method and Unsupervised Feature Learning." Paper presented at IEEE International Conference on Systems, Man, and Cybernetics.
6. Z. Li, S.E. Li, R. Li, B. Cheng, and J. Shi (2017) "Online Detection of Driver Fatigue Using Steering Wheel Angles for Real Driving Conditions." *Sensors*, 17(3): 495.
7. A.G. Correa, L. Orosco, and E. Laciar (2014) "Automatic Detection of Drowsiness in EEG Records Based on Multimodal Analysis." *Medical Engineering and Physics*, 36: 244–249.
8. R. Pooneh, R. Tabrizi, and A. Zoroofi (2009) "Drowsiness Detection Based on Brightness and Numeral Features of Eye Image." Paper presented at Fifth International Conference on Intelligent information Hiding and Multimedia Signal Processing.

Chapter 11

Prediction of Gastric Cancer from Gene Expression Dataset Using Supervised Machine Learning Models

P. Manikandan, D. Ruban Christoper,
and Luthuful Haq

11.1 Introduction

The geographical incidence of gastric cancer has changed dramatically since the 1940s. Gastric cancer was the most common reason behind cancer death in men. Recent studies reported that gastric cancer is 2.2 times more likely to be diagnosed in males compared with females (Rawla and Barsouk, 2019). Mortality rates of gastric cancer in males are high in several countries, such as Central and East Asia and Latin America. Gastric cancer peaks within the seventh decade of life. Often, a delay in diagnosing can account for the poor prognosis. Luckily, dedicated analysis into its pathological process and identification of recent risk factors, treatment and advanced endoscopic techniques offer aid to earlier detection of gastric cancer. Recognition that *Helicobacter pylori* infection causes most stomach ulcers has revolutionized the approach to gastric cancer these days.

Gastric cancer is comprised of two pathological variants: intestinal and diffuse. The intestinal type is the end result of an inflammatory condition that progresses from inflammation to symptom gastritis and finally to intestinal metaplasia and abnormal condition. This kind is common among older men, in contrast to the diffuse sort, that is rife among girls and in people below the age of 50. The diffuse type, characterized by the occurrence of *linitis plastica*, is related to an unfavorable prognosis, as a result of which the diagnosis is commonly delayed until the illness is advanced. Gastric *H. pylori* infection is closely related to this type, as in the intestinal type.

The XGBoost ensemble method is used to classify patients at high or low risk of developing gastric cancer. The dataset was collected from a single facility in Japan from 2006 to 2017. The results showed that XGBOOST performed best among 10 classification models built (Taninaga et al., 2019). The basic principles, advantages and disadvantages, training, and testing of data with the focus on the application of Machine Learning techniques used in the diagnosis of gastric cancer were studied (Danish et al., 2021). Extreme gradient boosting (XGBoost) had the best performance in predicting cancer development (AUC 0.97, 95%CI 0.96–0.98), and was superior to conventional logistic regression (AUC 0.90, 95% CI 0.84–0.92). With the XGBoost model, the number of patients considered at high risk of gastric cancer was 6.6 percent, with a miss rate of 1.9 percent. A total of 26 clinical variables were input into these models (Leung et al., 2021). A gradient boosting decision tree (GBDT) is a type of ensemble learning technique that can be used as a predictive model for the diagnosis of gastric cancer, based on noninvasive characteristics. The dataset consists of 709 samples and was collected from Zhejiang Provincial People's Hospital. The accuracy of the GBDT model was 83 percent (Zhu et al., 2020).

The Computerized Decision Support (CDS) system works as an assistant to doctors in the field of gastroenterology. The CDS system helps in identifying the cancerous area in the endoscopic images of the scaffold, to make a better diagnosis. This is helpful in determining the area and for biopsy samples taken from the gastric cancer patient (Yasar et al., 2019). A nested case-control study is performed using gastric cancer data from the Korean National Cancer Screening Program. The dataset consists of 16,584,283 Korean men and women, aged 40 years and older. Logistic regression model achieved the highest confidence of 95 percent (Jun et al., 2017). Identification and treatment in the early stage can significantly improve the prognosis of gastric cancer. Plasma samples were collected from 15 Early Gastric Cancer (EGC) patients and 15 healthy controls. As a result, the logistic regression and orthogonal signal correlation-partial least square discriminant analysis (OPLS-DA) model helps in identifying the plasma proteomics that could help distinguish EGC patients from healthy controls (Zhou et al., 2019). The relationship between Neutrophil-to-Lymphocyte (NLR) and tumor aggressiveness in gastric cancer patients was determined and investigated through the prognostic significance of NLR. Univariate analysis and variations of this parameter were also shown to be correlated with tumor progression. NLR values should be considered a useful follow-up parameter (Sahin et al., 2017). The cancer-related mortality and Potential Years of Life Lost (PYLL) were calculated by age, sex, districts (urban or rural), to describe the patterns of life lost to cancers. The dataset consists of 255 registries that were incorporated into the registration office in 2013, covering 226,494,490 people from urban areas. Three categories of people were found – those in urban areas, males and people over 60 years – who were suffering from more serious cancer deaths and life lost (Yan et al., 2019).

Cancer incidence and mortality produced by the International Agency for Research on Cancer on Worldwide were estimated. Some 19.3 million new cancer cases and almost 10.0 million cancer deaths occurred in 2020. The global cancer burden is expected to be 28.4 million cases in 2040, a 47 percent rise from 2020 (Sung et al., 2021). Clinicopathological factors predictive of lymph node metastasis in patients were identified with the poorly differentiated early gastric cancer. Some 1,005 patients were included in the analysis. The machine learning model, named Logistic regression, revealed that lymph node metastasis was significantly associated with sex, tumor size, depth of tumor invasion, and lymphatic involvement (Lee et al., 2012). The potential relationship between the severity of inflammation and prognosis in cancer patients was investigated. Patients' records of 220 gastric surgery patients, from January 2002 to December 2006, were analyzed with the univariate analysis (Lee et al., 2013). Medical check-up data of 25,942 participants from 2006 to 2017 at a single facility in Japan was analyzed using machine learning techniques. The results showed that XGBoost outperformed logistic regression and showed the highest area under the curve value as 0.899 (Taninaga et al., 2019). A microRNA panel was identified in the serum of patients to predict gastric cancer non-invasively with high accuracy and sensitivity. The Support Vector Machine Classifier outperformed the other models with the highest accuracy of 95 percent (Huang et al., 2018).

The most important aspects of gastric cancers, which include epidemiology, risk factors, classification, diagnosis, prevention, and treatment, were briefly reviewed (Sitarz et al., 2018). Supervised learning methods were used to distinguish the three gastric cancer subtypes. Leave-one-out cross-validation error was 0.14, suggesting that more than 85 percent of samples were classified correctly (Shah et al., 2011). New classification of gastric cancers and the up-to-date guidance in the application of molecular testing were reviewed. Recent advances in molecular medicine have not only shed light on the carcinogenesis of gastric cancer, but also offered novel approaches regarding prevention, diagnosis, and therapeutic intervention (Hu et al., 2012). The current applications of convolutional neural networks (CNN) in identification of gastric cancer were reviewed. The transfer learning approaches in CNN, such as AlexNet, ResNet, VGG, Inception and DenseNet, were used to identify gastric cancer. A total of 27 articles were retrieved for the identification of gastric cancer using medical images (Zhao et al., 2022). Deep convolutional neural networks (CNNs) are applied to automatically classify Magnifying-Narrow Band Imaging (M-NBI) images into two groups: normal gastric images and EGC images. VGG16, InceptionV3 and InceptionResNetV2 are selected to classify the M-NBI image. Experimental results show that transfer learning of deep CNN features performs better than traditional handcraft methods (Liu et al., 2018). Hence, it is necessary to develop an accurate and rapid screening method for diagnosis of gastric cancer. Recent works suggests that the Artificial Intelligence and Big Data technologies are effective in improving the screening, prediction and diagnosis in the medical fields (Hinton et al., 2017; Nishio et al., 2018). Based on the existing literatures, this research work also focused on classifying the gastric cancer dataset using supervised machine learning models. The rest of this chapter is organized as follows. Section 11.2 describes the Methodology and Section 11.3 shows the experimental results on gastric cancer dataset and discusses the outcomes. Finally, the conclusion and future enhancement are given in Section 11.4.

11.2 Methodology

This research work compares four supervised machine learning algorithms such as Logistic Regression (LR), Decision Tree (DT), Naïve Bayes (NB), and K-Nearest Neighbor (KNN) to predict the best classification method for classification of gastric cancer gene expression dataset. The overall framework for this research work is shown in Figure 11.1.

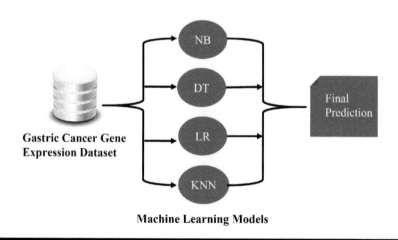

Figure 11.1   Overall framework of this research work.

11.2.1 Dataset Description

The gastric cancer gene expression dataset was collected from Gene Expression Omnibus with the GEO accession code (GSE2685) which consists of 4,524 attributes that include the records of 30 patients (Hippo et al., 2002). The python programming language is used to analyze the performance of the supervised machine learning algorithms.

11.2.2 Classification Techniques

In order to predict the outcome of the dataset, the algorithm processes a training set containing a set of attributes and the respective outcome, usually called target or prediction attribute. In this work, the supervised machine learning algorithms predict which of the algorithm is most suitable for classifying the gastric cancer gene expression dataset. The supervised machine learning algorithms namely Logistic Regression (LR), Decision Tree (DT), Naïve Bayes (NB) and K-Nearest Neighbor (KNN) are used to find out which fits effectively for the gastric cancer gene expression dataset.

11.2.2.1 Logistic Regression (LR)

Logistic Regression is basically a supervised machine learning algorithm and it is used to determine the relationship between the dependent variable (x) to the set of independent variables (y). In a classification problem the target variable or output (y) can only be a discrete variable for the given set of features (x). The model builds the regression model to predict the probability that the given data belongs to the category numbered as 1. Just like linear regression, it follows the linear algorithm. Logistic regression data models use the sigmoid function (Eq. 11.1) to convert the values within the range of 0–1 with the threshold value of 0.5 (Brindha et al., 2021).

$$\text{Sigmoid Function } S(x) = \frac{1}{1+e^{-x}} \tag{11.1}$$

11.2.2.2 Decision Tree (DT)

The decision tree is a non-parametric supervised machine learning algorithm (Su et al., 2007) that can be used for both classification and regression problems. But mostly it is employed in classification problems. It is a flowchart with a tree-like structure where the internal node represents the feature, the branch represents the decision rule, and the leaf node represents the outcome. Decision tree works on the sum of product form which is known as the disjunctive normal form. It has two popular attribute selection measures, such as the Entropy and the Gini Index. Entropy (Eq. 11.2) is used to measure the impurity of the gastric cancer gene expression dataset. The Gini Index (Eq. 11.3) computes the degree of probability of a specific variable that has been wrongly classified. It works on categorical variables and gives the outcome of success or failure:

$$\text{Entropy} = \sum_{i=1}^{n} -p\left(C_i\right)\log_2\left(p\left(C_i\right)\right) \tag{11.2}$$

$$Gini = 1 - \sum_{i=1}^{n} -p^2\left(C_i\right) \tag{11.3}$$

11.2.2.3 Naïve Bayes (NB)

Naïve Bayes (NB) is a probabilistic machine learning algorithm based on the Bayes Theorem that can be used in a wide variety of classification tasks. It is called Naïve because it assumes that the occurrence of a certain feature is independent of the occurrences of the features. It is simple and easy to us and it doesn't require as much training data and it handles both continuous and categorical data. NB (Eq. 11.4) is one of the simplest and most effective classification algorithms which helps in building the fast machine learning models that can make quick predictions (Mortezagholi et al., 2019). It leads to a linear decision boundary in many common cases.

Bayes depends on the principle of the Bayes theorem.

$$P(A|B) = P(B|A) * P(A) / P(B) \tag{11.4}$$

where:
P(A|B) is posterior probability
P(B|A) is likelihood probability
P(A) is prior probability
P(B) is marginal probability

11.2.2.4 K-Nearest Neighbor (KNN)

K-Nearest Neighbor (KNN) is a non-parametric supervised learning algorithm which uses proximity to make classifications or predictions about grouping of an individual data point (Li et al., 2012). KNN works by finding the distance between a query and all the samples in the data. In the KNN algorithm, the K value is initialized to choose the number of neighbors. Then, the distance between the current objects and the k cluster centroids is calculated. Assign the current sample to that cluster to which it is closest. Compute the "cluster centers" of each cluster. These become the new cluster centroids. These steps have to be repeated, until the convergence criterion is satisfied.

11.3 Results and Discussion

In this research work, the experimental measures are calculated by using the performance factors such as the classification Accuracy, Precision, Recall and F1-Score to determine the best algorithm for the gene expression dataset. The accuracy measure and performance metric values for the gastric cancer gene expression dataset are depicted in Table 11.1. In this research work, the basic exploration for the gastric cancer gene expression dataset was performed like checking for null values, duplicate values and data types etc.

Table 11.1   Comparison of Performance Metrics for Various Supervised Machine Learning Algorithms Using the Gastric Cancer Gene Expression Dataset

Models	Accuracy	Precision	Recall	F1 score
LR	0.625	0.55	0.62	0.58
DT	0.875	0.94	0.88	0.89
NB	0.75	0.88	0.75	0.80
KNN	0.5	0.40	0.50	0.44

Of the 4,524 features present in the dataset, "class" is the dependent variable with three classes "Normal," "Diffuse," and "Intestinal." Using the countplot from the seaborn library, the count of three classes present in the dataset was predicted (Figure 11.2).

['AB006782_at', 'AC002077_at', 'D26129_at', 'D38535_at', 'D42047_at', 'D78134_at', 'D87742_at', 'HG3162-HT3339_at', 'J03915_s_at', 'J04982_at', 'J05412_at', 'L07590_at', 'L13744_at', 'L32179_at', 'L41668_rna1_at', 'L76465_at', 'M12759_at', 'M21812_at', 'M31328_at', 'M61855_at', 'M62628_s_at', 'M63154_at', 'M74542_at', 'S68616_at', 'S69232_at', 'U05259_rna1_at', 'U15177_at', 'U19948_at', 'U21931_at', 'U26424_at', 'U27325_s_at', 'U57094_at', 'U59752_at', 'U66052_at', 'U66702_at', 'U70663_at', 'U79288_at', 'U83461_at', 'X05409_at', 'X05997_at', 'X51698_s_at', 'X52003_at', 'X53961_at', 'X58529_at', 'X66839_at', 'X69920_s_at', 'X76223_s_at', 'X76342_at', 'X96752_at', 'Z29574_at']

To find the correlation of 4,522 independent variables in the dataset, the Pearson's r function is used from the scipy.stats module and converts the independent variables to integer datatype. A loop is used with a range of 0–4521 to find the correlation between the features and dependent variable. Since the dependent variable is a

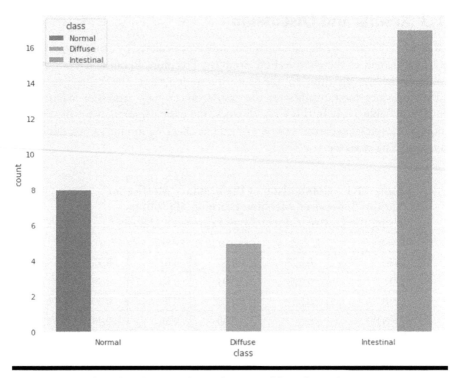

Figure 11.2   Class distribution (dependent variable) of gastric cancer gene expression dataset.

['AB006782_at', 'AC002077_at', 'D26129_at', 'D38535_at', 'D42047_at', 'D78134_at', 'D87742_at', 'HG3162-HT3339_at', 'J03915_s_at', 'J04982_at', 'J05412_at', 'L07590_at', 'L13744_at', 'L32179_at', 'L41668_rna1_at', 'L76465_at', 'M12759_at', 'M21812_at', 'M31328_at', 'M61855_at', 'M62628_s_at', 'M63154_at', 'M74542_at', 'S68616_at', 'S69232_at', 'U05259_rna1_at', 'U15177_at', 'U19948_at', 'U21931_at', 'U26424_at', 'U27325_s_at', 'U57094_at', 'U59752_at', 'U66052_at', 'U66702_at', 'U70663_at', 'U79288_at', 'U83461_at', 'X05409_at', 'X05997_at', 'X51698_s_at', 'X52003_at', 'X53961_at', 'X58529_at', 'X66839_at', 'X69920_s_at', 'X76223_s_at', 'X76342_at', 'X96752_at', 'Z29574_at']

Figure 11.3   List of top 50 features selected by ANOVA method.

categorical variable, the label encoder is used to convert the values to numeric. Since the number of features is huge, many of the features do not provide much meaning relevant to the study. Hence, the Feature Selection is employed using Analysis of Variance (ANOVA) to select the top 50 features which will be used to model (Figure 11.3). The top 50 features selected by the ANOVA method are passed as an independent feature to the four supervised machine learning classification models such as LR, DT, NB, and KNN. The performance metrics, namely Accuracy,

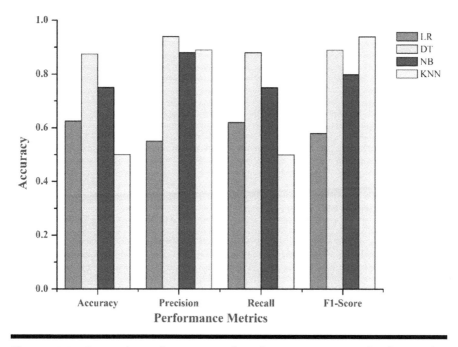

Figure 11.4 **Comparison of accuracy and performance measures for the gastric cancer gene expression dataset.**

Precision, Recall and F1-Score, are used to evaluate the performance of the machine learning models. From the experimental results (Figure 11.4), it is observed that the Decision Tree (DT) algorithm performs better than the other supervised machine learning algorithms. Out of 30 patients, 5 patients are identified to have diffuse gastric cancer, 17 were found to have intestinal gastric cancer, and 8 are found to be normal not have gastric cancer.

11.4 Conclusion

In this research work, the performance of four supervised machine learning algorithms, namely Logistic Regression, Decision Tree, Naïve Bayes and K-Nearest Neighbor, were analyzed. The gastric cancer gene expression dataset was used to calculate the performance by using the train-test split method. And, finally, this research has analyzed the algorithms using the Accuracy and Performance Metrics. From the results, it is observed that the DT algorithm performs better than the other algorithms. Out of 30 patients, 5 patients are identified to have diffuse Gastric Cancer, 17 were found to have Intestinal Gastric Cancer, 8 are found to be NORMAL not have gastric cancer. In future, the DT algorithm can be experimented on other datasets also. And, in future, the DT algorithm should be modified to obtain more effective results. Also, the classification algorithms can be analyzed using other parameters such as the cross-validation split.

References

Brindha, Senthil Kumar, et al. (2021) "Logistic Regression for Gastric Cancer Classification Using Epidemiological Risk Factors in Cases and Controls." *Science & Technology Journal*, 9: 154–160.

Danish, Jamil, Sellappan, Palaniappan, Asiah, Lokman, et al. (2021) "Diagnosis of Gastric Cancer Using Machine Learning Techniques in the Healthcare Sector: A Survey." *Informatica*. 45(7): 147–166.

Hinton, D. J. et al. (2017) "Metabolomics Biomarkers to Predict Acamprosate Treatment Response in Alcohol-Dependent Subjects." *Scientific Reports*. 7: 2496. PMID: 28566752

Hippo, Y., Taniguchi, H., Tsutsumi, S., Machida, N. et al. (2002) "Global Gene Expression Analysis of Gastric Cancer by Oligonucleotide Microarrays." *Cancer Research*, 62(1): 233–240.

Hu, B., El Hajj, N., Sittler, S., Lammert, N., Barnes, R., and Meloni-Ehrig, A. (2012) "Gastric Cancer: Classification, Histology and Application of Molecular Pathology." *Journal of Gastrointestinal Oncology*, 3(3): 251–261.

Huang, Y., Zhu, J., Li, W., Zhang, Z., Xiong, P., Wang, H., and Zhang, J. (2018) "Serum MicroRNA Panel Excavated by Machine Learning as a Potential Biomarker for the Detection of Gastric Cancer." *Oncology Reports*, 39(3): 1338–1346.

Jun, J.K., Choi, K.S., Lee, H.Y., Suh, M., Park, B., Song, S.H., Jung, K.W., Lee, C.W., Choi, I.J., Park, E.C., and Lee, D. (2017) "Effectiveness of the Korean National Cancer Screening Program in Reducing Gastric Cancer Mortality." *Gastroenterology*, 7152(6): 1319–1328.e7.

Lee, D.Y., Hong, S.W., Chang, Y.G., Lee, W.Y., and Lee, B. (2013) "Clinical Significance of Preoperative Inflammatory Parameters in Gastric Cancer Patients." *Journal of Gastric Cancer*, 13(2): 111–116.

Lee, J.H., Choi, M.G., Min, B.H., Noh, J.H., Sohn, T.S., Bae, J.M., and Kim, S. (2012) "Predictive Factors for Lymph Node Metastasis in Patients with Poorly Differentiated Early Gastric Cancer." *British Journal of Surgery*, 99(12): 1688–1692.

Leung, W.K., Cheung, K.S., Li, B., Law, S.Y.K., and Lui, T.K.L. (2021) "Applications of Machine Learning Models in the Prediction of Gastric Cancer Risk in Patients After *Helicobacter Pylori* Eradication." *Alimentary Pharmacology and Therapeutics*, 53(8): 864–872.

Li, C., Zhang, S., Zhang, H., Pang, L., Lam, K., Hui, C., Zhang, S. (2012) "Using the K-Nearest Neighbor Algorithm for the Classification of Lymph Node Metastasis in Gastric Cancer." *Computational and Mathematical Methods in Medicine*. 2012: 876545.

Liu, X., Wang, C., Hu, Y., Zeng Z., Bai J., and Liao G. (2018) "Transfer Learning with Convolutional Neural Network for Early Gastric Cancer Classification on Magnifying Narrow-Band Imaging Images." Paper presented at 2018 25th IEEE International Conference on Image Processing (ICIP), pp. 1388–1392.

Mortezagholi, A., Khosravizadeh, O., Menhaj, M.B., Shafigh, Y., and Kalhor, R. (2019) "Make Intelligent Gastric Cancer Diagnosis Error in Qazvin's Medical Centers: Using Data Mining Method." *Asian Pacific Journal of Cancer Prevention*, 20(9): 2607–2610.

Nishio, M. et al. (2018) "Computer-Aided Diagnosis of Lung Nodule Using Gradient Tree Boosting and Bayesian Optimization." *PLoS One* 13, e0195875. PMID: 29672639.

Rawla, P., and Barsouk, A. (2019) "Epidemiology of Gastric Cancer: Global Trends, Risk Factors and Prevention." *Prz Gastroenterology*, 14(1): 26–38.

Sahin, A.G., Aydin, C., Unver, M., and Pehlivanoglu, K. (2017) "Predictive Value of Preoperative Neutrophil Lymphocyte Ratio in Determining the Stage of Gastric Tumor." *Medical Science Monitor*, 23: 1973–1979.

Shah, M.A., Khanin, R., Tang, L., Janjigian, Y.Y., Klimstra, D.S., Gerdes, H., and Kelsen, D.P. (2011) "Molecular Classification of Gastric Cancer: A New Paradigm." *Clinical Cancer Research*, 17(9): 2693–2701.

Sitarz, R., Skierucha, M., Mielko, J., and Offerhaus, G.J.A., Maciejewski, R., and Polkowski, W.P. (2018) "Gastric Cancer: Epidemiology, Prevention, Classification, and Treatment." *Cancer Management and Research*, 10: 239–248

Su, Y, Shen, J, Qian, H, Ma, H, Ji, J, Ma, H, Ma, L, … Shou, C. (2007) "Diagnosis of Gastric Cancer Using Decision Tree Classification of Mass Spectral Data." *Cancer Science*, 98(1): 37–43.

Sung, H., Ferlay, J., Siegel, R.L., Laversanne, M., Soerjomataram, I., Jemal, A., and Bray, F. (2021) "Global Cancer Statistics 2020: GLOBOCAN Estimates of Incidence and Mortality Worldwide for 36 Cancers in 185 Countries." *CA: A Cancer Journal for Clinicians*, 71(3): 209–249.

Taninaga, J., Nishiyama, Y., Fujibayashi, K. et al. (2019) "Prediction of Future Gastric Cancer Risk Using a Machine Learning Algorithm and Comprehensive Medical Check-Up Data: A Case-Control Study." *Scientific Reports*, 9: 12384.

Yan, Y, Chen, Y, Jia, H, Liu, J, Ding, Y, Wang, H, Hu, Y, … Li, S. (2019) "Patterns of Life Lost to Cancers with High Risk of Death in China." *International Journal of Environmental Research and Public Health*, 16(12): 2175.

Yasar, A., Saritas, I., Korkmaz H. (2019) "Computer-Aided Diagnosis System for Detection of Stomach Cancer with Image Processing Techniques." *Journal of Medical Systems*, 43(4): 99.

Zhao, Y., Hu, B., Wang, Y., Yin, X., Jiang, Y., and Zhu, X. (2022) "Identification of Gastric Cancer with Convolutional Neural Networks: A Systematic Review." *Multimedia Tools and Applications*, 81(8): 11717–11736.

Zhou, B., Zhou, Z., Chen, Y., Deng, H., Cai, Y., Rao, X., Yin, Y., and Rong, L. (2020) "Plasma Proteomics-Based Identification of Novel Biomarkers in Early Gastric Cancer." *Clinical Biochemistry*, 76: 5–10.

Zhu, S.L., Dong, J., Zhang, C., Huang, Y.B., and Pan, W. (2020) "Application of Machine Learning in the Diagnosis of Gastric Cancer Based on Noninvasive Characteristics." *PLoS One*, 5(12): e0244869.

Chapter 12

Sewer Pipe Defect Detection in CCTV Images Using Deep Learning Techniques

P. L. Chithra and P. Bhavani

12.1 Introduction

The sewer pipe systems are a component of civil infrastructure that is designed to collect wastewater. In the United States, 800,000 miles of public and 500,000 miles of private lateral sewers are connected to public sewer pipes. Most of the sewer systems are between 30 and 100 years old and suffer pipe blockage, which is mainly caused by grease and debris. Sewer defects are caused by root intrusion, line breaks, and unequal amounts of inflow and infiltration. For example, cracks can be divided into longitudinal cracks, vertical cracks and complex cracks consisting of both longitudinal and vertical ones. Pipe deposits are of different types, such as attached deposits (attached to the pipe wall) and settled deposits (settled on the pipe floor). As for tree root intrusion, this can belong to the mass root (high density or large occupying area) or minor root (scattered and with low density). All these defects such as cracks, deposits, tree root intrusions, as well as water infiltrations, impair the performance of the sewer pipe. Currently, visual inspection techniques such as closed-circuit television (CCTV) are commonly used for underground sewer pipe inspection.

A CCTV usually consists of a camera and an illumination device mounted on a tractor. During the inspection, the CCTV unit moves along the interior pipe wall and transmits the inspection video to an external monitor on the ground. While encountering a pipe defect or pipe lateral, the inspector stops the unit and zooms the camera into the abnormal part to check whether there are potential defects or not. After the inspection, the inspector watches the captured images or videos to identify the defect type and location. Such manual interpretation of the inspection images or videos is time-consuming, labor-intensive and the results can be subjectively inaccurate. For this reason, an efficient computer vision technique has been used to interpret the inspection images or videos automatically. However, conventional computer vision techniques require the design of complex feature extractors and the images used for training need a large amount of pre-processing.

In addition, the training process is tedious and inefficient. In recent years, deep learning has achieved promising performance in various computer vision tasks such as image classification and object detection. Compared with conventional computer vision techniques, approaches based on deep learning are capable of extracting image features automatically and there is no requirement for image pre-processing, which greatly improves the accuracy and efficiency. Therefore, an automated defect detection approach is proposed in this study, based on faster region-based convolutional neural network (Faster R-CNN), which is a deep learning model for object detection and also for identifying and locating sewer pipe defects from CCTV images.

The rest of the chapter is organized as follows, Section 12.2 contains the existing works related to sewer pipe defects, Section 12.3 explains the various methodologies, Section 12.4 presents the experimental results and, finally, Section 12.5 yields the conclusion.

12.2 Related Work

One of the challenging problems in today's world is the detection of automatic sewer pipes. The sewer pipe system is designed to collect sewage water and groundwater or stormwater [1], which are monitored by CCTV videos in some municipalities [2, 3]. It is expected that approximately 19,500 sewer pipe systems [4] can handle the flow of 50 billion gallons of raw sewage but most of the sewer systems have become obsolete, as they are 30–100 years old [5]. There are various sewer pipe defects such as cracks, roots, deposits, and infiltrations, etc. [6]. For example, cracks [7, 8] are divided into longitudinal cracks (same orientation of the pipeline), vertical cracks (orientation is vertical to the pipeline) as well as complex cracks consisting of both longitudinal and vertical ones [9].

Deposits are of different types, such as attached deposits (attached to the pipe wall) and settled deposits (settled on the pipe ground) [10]. As for the tree root intrusion, this can be a mass root intrusion (high density or large occupying area) or a minor root (scattered and with low density). In the visual inspection technique by CCTV video images, a large number of inspection images are mostly conducted manually to identify defect types and locations, which is time-consuming, costly, labor-intensive, and inaccurate. Computer vision is performed by a computer or machine that can recognize digital images or videos [11]. The computer vision techniques [12, 13] are Haar detector [14], SIFT [15], HOG [16] using larger HOG filters [17] and the deformable part-based model (DPM) [18]. However, these methods require the complex manual design of the feature extractor and training process, which are inefficient. The digital image processing techniques are segmentation [19], histogram equalization, and transformation.

Deep learning-based approaches have been developed for object detection [20] to address the limitation of conventional methods. The R-CNN method uses region proposals generated through an external method called the selective search for the input image. Each warped region proposal image is forwarded to a CNN model to compute the features into a support vector machine (SVM) [4] classifier to calculate the classification scores. Bounding box regression is then conducted for the classified image so that the location of each object can be predicted. One limitation of R-CNN is that the multi-staged training process is time-consuming and requires large computation cost. In addition, the detection speed is quite slow for each image as the convolution, classification, and regression need to be implemented for each region.

The fast R-CNN is applied to the input image to produce the feature map, on which the region proposals (RPN) [21] are generated by the selective search. Through a Region of Interest (ROI) pooling layer, the generated region proposals are converted into fixed-length vectors, which are fed into the fully connected layers for both classification and bounding box regression. Compared with RCNN, fast R-CNN [22] performs training using multi-task loss in a single-stage manner, which greatly reduces the training time. The detection is faster because the convolution process is only conducted once for the original image instead of for all the region proposals, and the detection accuracy speed is improved. Cheng et al. have adopted a method to identify cracks depending on conventional computer vision technique using faster R-CNN with deep learning techniques [6]. The detection model is trained using 3,000 images collected from CCTV inspection videos of sewer pipes.

After training, the model is evaluated in terms of detection accuracy and computation cost using mean average precision (MAP), missing rate, detection speed, and training time. Specifically, the increase of dataset size and convolutional layers can improve the model's accuracy. The experiment results demonstrate that dataset size, initialization network type, training mode and network hyper-parameters model performance are improved. The adjustment of hyper-parameters such as filter dimensions or stride values contributes to higher detection accuracy, achieving MAP of 83 percent accuracy and faster detecting speed in the sewer pipe. It is very efficient for the automated detection of sewer pipe defects in CCTV.

A robust methodology for detecting faults in sewer pipes has been developed. Faults of all defined type are detected in still images (81.5 percent accuracy). The methodology can be effectively applied to continuous footage (80 percent accuracy). Random Forests are identified as the most appropriate machine learning classifier. Smoothing greatly reduces the False Positive Rate [22].

This method presents the development of an automated tool to detect some defects such as cracks, deformation, settled deposits, and joint displacement in sewer pipelines. The automated approach is dependent upon using image-processing techniques and several mathematical formulas to analyze output data from CCTV camera images [23]. Automated CCTV interpretation could improve the speed and accuracy of inspections. The defects in CCTV inspection use CNNs for classifying sewer defects. The CNNs were trained and tested on a set of 12,000 images collected from over 200 pipelines [24].

12.3 Proposed Methodologies

The deep learning technique for object detection using faster R-CNN with the combinations of different CNN architecture networks and image processing techniques is used to detect defects.

12.3.1 Faster R-CNN for Object Detection

The faster R-CNN is faster than R-CNN. It is used to predict the region proposals that are then reshaped using a Region of Interest (ROI) pooling layer which is used to classify the image within the proposed region and predict the offset values for the bounding boxes.

12.3.1.1 Convolutional Layer

Every neuron takes inputs from a rectangular n*n section of the previous layer. The rectangular section is called a local receptive field. The layer is combined with bias and weights to produce a feature map. A parameter could be viewed as a trainable filter or kernel F, the convolutional process could be considered as acting as an image convolution and it takes input from the previous layer (Eq. 12.1). Sometimes it may be called a trainable filter from the input layer to the hidden layer, based on the feature map with shared weights and bias.

$$xij^{(l)} = \sigma(b + \sum_{c=0}^{n} Wr,c\ X_{i+r,j+c}^{(l-1)}) \tag{12.1}$$

12.3.1.2 Feature Map

The convolved feature matrix is formed by sliding the filter over the image and computing the dot product. Every layer feature map output acts as input to the next layer.

12.3.1.3 Region Proposal Network (RPN)

In RPN training, each region proposal is labeled to indicate whether an object has a defective region or not. If the intersection of union (IOU) between the proposal and the ground-truth box exceeds the threshold (0.7 in this study), or if it has the highest IOU with a ground-truth box, the proposal is assigned as a positive training sample, if the maximum IOU between a non-positive proposal and the ground-truth box is lower than the corresponding threshold (0.3 in this study), the proposal is regarded as a negative sample. RPN in faster R-CNN is trained using a multi-task loss function calculated by Eq. 12.2.

$$L\left(\{p_i\},\{t_i\}\right) = \frac{1}{N_{cls}}\sum_i L_{cls(p_i,p_i^*)} + \frac{1}{N_{reg}}\sum_i p_i^* L_{reg(t_i,t_i^*)} \qquad (12.2)$$

where 1 indicates the number of the anchor, p_i represents the predicted probability of anchor i being one type of sewer pipe defect. p_i * represents the ground-truth label of anchor i, where p_i * equals 1 if the anchor is a positive sample and 0 if it is negative. t_i is a vector indicating the coordinates of the predicted bounding box while ti* represents the ground-truth bounding box related to a positive anchor. N_{cls} and N_{reg} are two normalization factors for the two components and are weighted by a balancing parameter λ. With reference to a related study [22], the value of N_{cls} is set to be 256 and N_{reg} equals 2,400 which is an approximate value of the number of anchor locations in the model. The balancing parameter λ is set to 10, so that both cls and reg terms are roughly weighted equal.

12.3.1.4 Region of Interest (ROI) Pooling

ROI operation is widely used in object detection tasks to reuse the feature map from CNN (Figure 12.1). It allows training object detection system to be used in an end-to-end manner and significantly speeds up both training and testing time.

12.3.1.5 Max Pooling Layer

Max pooling is a sample based on the discretization process. The main objective is to download a sample of an input representation of an image to hidden layer output,

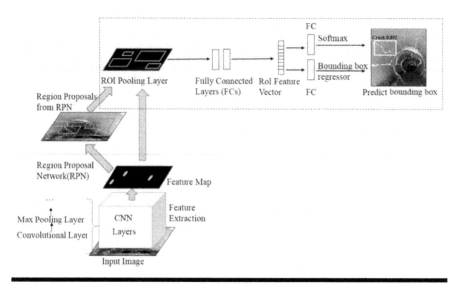

Figure 12.1 Architecture model for developed faster R-CNN for sewer pipe defect detection.

1	3	2	1
2	9	1	1
1	3	2	3
5	6	1	2

9	2
6	3

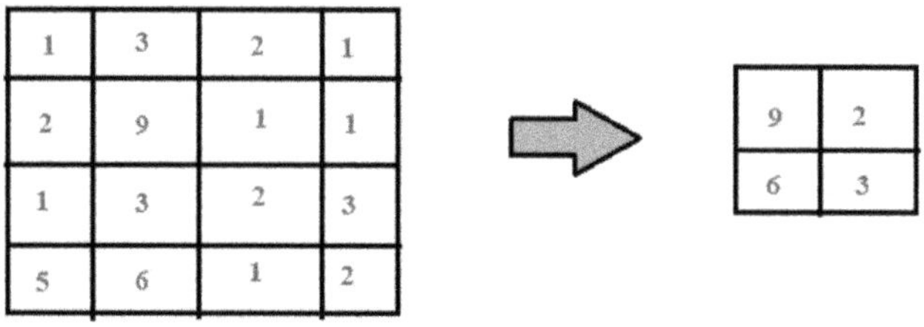

Figure 12.2 Max pooling conversion process.

etc. (Figure 12.2). It reduces dimensionality and allows assumptions to be made about features contained in the sub-regions binned.

Max pooling uses the maximum value from each of a cluster of neurons at the prior layer and the hyperparameter values, filter (f) =2 stride(s) =2.

12.3.2 System Architecture

This proposed work is to detect sewer pipe defects trained in different networks and apply some background effects. Defects have been detected and feature matching is carried out. The flow diagram for pipe defect detection is shown in Figure 12.3. In this method, sewer pipe images have been collected from CCTV inspection videos.

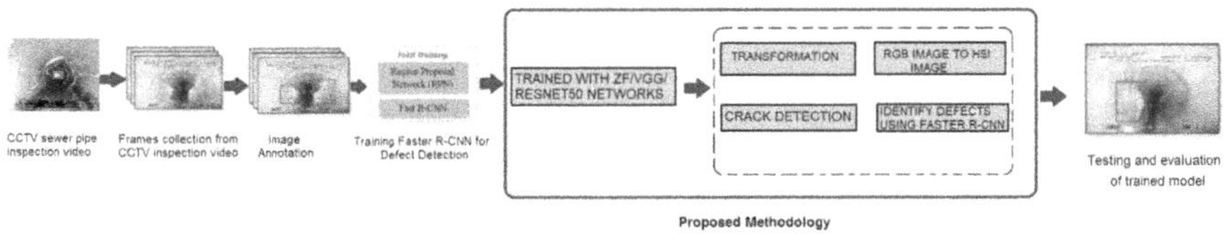

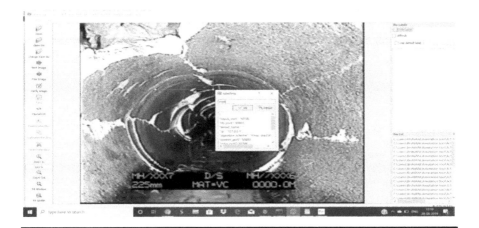

Figure 12.3   Proposed workflow.

Figure 12.4   Labeled input image.

12.3.2.1 Labeled Input Image

The CCTV inspection videos are converted into frames and these frames are considered the input images. Further, they have been annotated using a labeling tool. (Figure 12.4). Finally, bounding boxes are created to be saved as .xml extension, like VOC dataset.

12.3.2.2 Identify Defects Using Faster R-CNN

A faster R-CNN model consists of a region proposal network (RPN) and fast R-CNN detector. There are three main steps in applying faster R-CNN for object detection. First, a convolutional neural network is used on the raw images to extract features. A feature map of the raw image is generated through several layers of a CNN network, such as convolution, activation, pooling, and normalization. Based on the generated feature map, RPN, a single fully convolutional network generates region proposals with different aspect ratios and scales to distinguish foreground regions from background regions. In the third step, the generated region proposals are fed into the fast R-CNN detector for further refining, so that more accurate classification and bounding box localization are obtained. This predicts the location of the proposed bounding boxes regression, thereby getting the best nearby ground-truth box for the anchor box. The coordinates of the anchor are calculated using Eqs (12.3)–(12.6).

$$t_x = \frac{x - x_a}{w_a}, t_y = \frac{y - y_a}{h_a} \tag{12.3}$$

$$t_w = \log\frac{w}{w_a}, t_h = \log\frac{h}{h_a} \tag{12.4}$$

$$t_x^* = \frac{x^* - x_a}{w_a}, t_y^* = \frac{y^* - y_a}{h_a} \tag{12.5}$$

$$t_w^* = \log\frac{w^*}{w_a}, t_h^* = \log\frac{hh^*}{h_a} \tag{12.6}$$

12.3.2.3 Training with ZF/VGG/RESNET50 Networks

In this method, the faster R-CNN model is trained with different networks. First, the ZF network (without fully connected layers) is applied, using parameters like the number of filters, the dimension of each filter, and the stride values. For example, there are 96 convolution kernels with a dimension of 7 × 7 in the first convolutional layer and 256 kernels with a dimension of 7 × 7 in the second convolutional layer. Second is the VGG16 network with 3 × 3 convolutions filters. The ResNet50 convolutional neural network is trained on more than a million images from the ImageNet database and the network is 50 layers deep and can classify images into 1,000–2,000 object categories, such as crack, root, deposit, and infiltration and the input length is 6, it can apply four stride convolutions values and the filter size is 7, 3, 1, 1.

12.3.2.4 RGB Image Converted to HSV Image

The RGB color represents Red, Green and Yellow. An HSV is a color space in which H stands for hue, S stands for Saturation and V stands for Value. Hue is the measure of color; average pipe defects are perceived to have 200 different colors. Saturation is the relative purity of color, the brightness of an object is its intensity. Hue is defined as an angle: in 0 degrees is red, 120 degrees is green and 240 degrees is blue. The saturation value ranges from 0 to 1. Figure 12.5 shows the conversion of the RGB image to the HSV image. The angular relationship between tones around the color circle is easily identified. Equations (11.7)–(11.9) show Hue, Saturation and value, and produce the crack effectively.

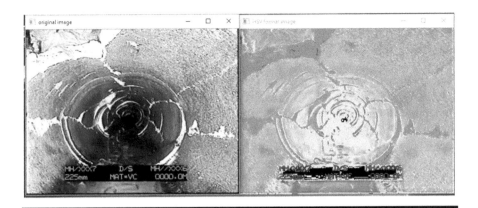

Figure 12.5   RGB image converted to HSV image.

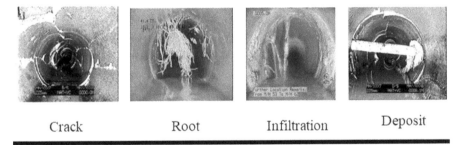

| Crack | Root | Infiltration | Deposit |

Figure 12.6   Crack detection.

$$V = R\left(\frac{1}{3}\right) + G\left(\frac{1}{3}\right) + B\left(\frac{1}{3}\right) \tag{12.7}$$

$$S = 1 - \frac{3}{(R+G+B)}\left[\min(R,G,B)\right] \tag{12.8}$$

$$H = \begin{cases} \theta & \text{if } B \leq G \\ 360 - \theta & \text{if } B > G \end{cases} \tag{12.9}$$

12.3.3 Crack Detection

Crack detection is used to identify cracks in an image and draw bounding boxes around them. This is a very important problem in computer vision in numerous applications. Figure 12.6 uses crack detection to identify crack roots, deposits, and infiltration.

12.4 Experimental Results

This proposed technique has been applied to detect defects, such as cracks, roots, deposits, infiltration, etc. This work takes 3,000 defects in various video images to prove experimental results. For defect detection, different sets of images are taken freely in different scenes, under various lighting conditions and they create defects in bounding boxes using faster R-CNN. Performance of the method is calculated using average precision (AP) and mean average precision (MAP) in overall defects and achieves better results. The proposed method applies some digital image processing techniques to produce better results. The dataset used in our experiments is a collection of sewer pipe inspection videos (CCTV footage), with that video converted into frames. The frames are more than 30,000–1,00,000 images depending upon CCTV videos.

12.4.1 Performance Analysis

The sewer pipe detects the defects such as cracks, roots, deposits, and infiltration performance of ZF networks. Figure 12.7 (a) shows average result, Figure 12.7 (b) shows the Resnet50 network gives better performance compared to ZF net, and Figure 12.7 (c) shows that VGG16 gives better performance than the ZF and Resnet50 networks.

Our experimental results show that the proposed model based on the VGG network outperforms well with a detection speed of 0.99 secs and training time of 8.573 secs better than the existing network of ZF and Resnet50. After training, this model is evaluated in terms of detection accuracy achieving MAP of 90 percent using the VGG network in the sewer pipe system. The defect detection MAP and the AP value are mentioned in Table 12.1.

12.4.2 Measure of Effectiveness

To calculate the AP and MAP for all the class, use Eqs. (12.10) and (12.11)

$$AP = \frac{1}{11} \sum_{r \in \{0,0,1,0,2,...1\}} \frac{\max p(\tilde{p}_r)}{\tilde{p}_r : \tilde{p}_r \geq r} \tag{12.10}$$

where $p(\tilde{\ }_r)$ is the measured precision at recall $\tilde{p}_r$

$$MAP = \frac{1}{N_{cls}} \sum_i AP_i \tag{12.11}$$

Table 12.1   Defect Detection Proposed by AP and MAP Values

Network	Datasets	MAP	AP			
			Crack	Root	Deposit	Infiltration
ZF	A (1000)	0.575	0.512	0.642	0.574	0.575
	B (2000)	0.701	0.624	0.798	0.698	0.687
	C (3000)	0.725	0.787	0.745	0.658	0.712
RESNET 50	A (1000)	0.780	0.749	0.846	0.827	0.699
	B (2000)	0.818	0.835	0.824	0.749	0.865
	C (3000)	0.923	0.923	0.889	0.979	0.902
VGG 16	A (1000)	0.961	0.957	0.989	0.928	0.971
	B (2000)	0.903	0.806	0.858	0.986	0.965
	C (3000)	0.843	0.898	0.789	0.899	0.789

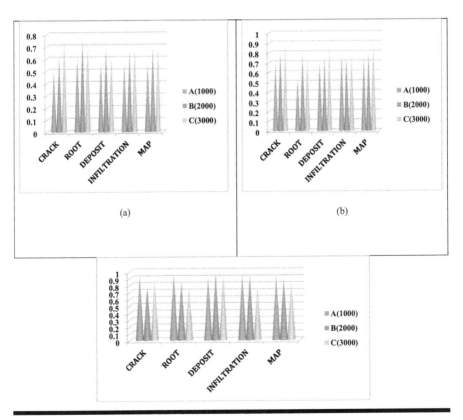

Figure 12.7   Chart for (a) ZF network, (b) RESNET50 network and (c) VGG 16 networks.

12.4.3 Experimental Outcome Results

The ZF network has a detection speed: for an oval crack it is 0.96, for a longitudinal crack it is 0.92, for an attached deposit it is 0.83, and for a vertical crack it is 0.85 seconds (Figure 12.8).

The Resnet 50 network has a detection speed: for an attached deposit it is 0.99 and 0.95, the mass root is 0.83, 0.98 and 0.93, a vertical crack is 0.93 seconds compared to ZF net (Figure 12.9).

The VGG16 network has a detection speed: for a longitudinal crack it is 0.99, 0.93, for infiltration it is 0.92 seconds compared to ZF net and Resnet50 (Figure 12.10).

Further, some digital image processing techniques such as region-based segmentation and cluster-based segmentation methods are applied in the output of the crack image (Figure 12.11). The transformations of horizontal flip, vertical flip, and their gray images are shown in Figure 12.12. Histogram equalizations for crack

Figure 12.8 Defect images trained with ZF network. (a) sample oval crack outcome, (b) longitudinal crack, settled and attached deposit (c) vertical crack.

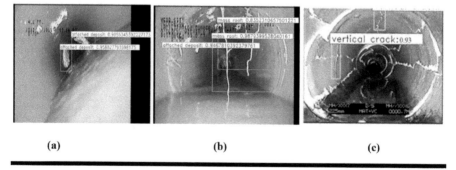

Figure 12.9 Defect images trained with Resnet50 network (a) attached deposit, (b) mass root (c) vertical crack.

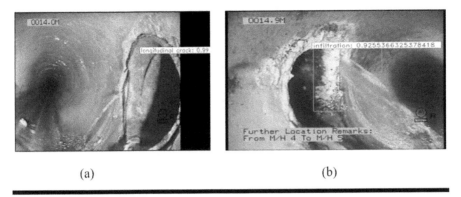

(a)   (b)

Figure 12.10 **Defect images trained with VGG16 network (a) longitudinal crack (b) infiltration.**

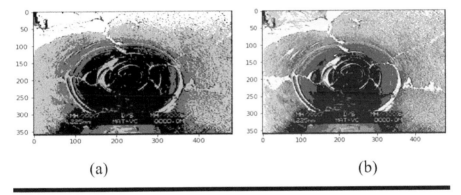

(a)   (b)

Figure 12.11 **(a) Region-based segments (b) image segmentation based on clustering.**

defected image are shown in Figure 12.13. Finally, segmented defects are created using bounding boxes (Figure 12.14).

12.5 Conclusion

This work suggests solutions to pipe system detection. Many pipe systems are old so various defects such as roots, cracks, deposits, and infiltration are found, which can lead to serious consequences. CCTV is used for sewer pipe inspection by capturing inspection videos. Manual interpretation of the CCTV inspection videos is time-consuming, and the assessment results are affected by the inspector's expertise and experience. Here defects are found quickly using faster R-CNN with VGG

Original Image

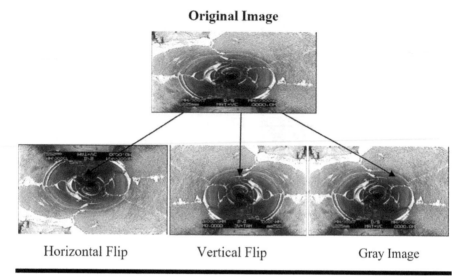

Horizontal Flip   Vertical Flip   Gray Image

Figure 12.12   Transformation for crack detection.

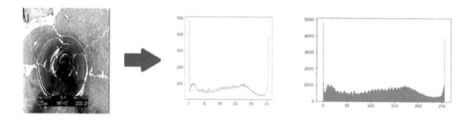

Figure 12.13   Histogram for crack detection.

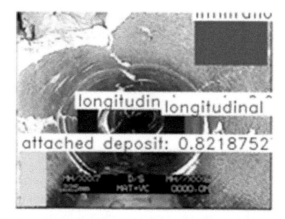

Figure 12.14   Segments the defects with bounding box.

techniques and some digital image processing techniques were applied, which show defects accurately and fix the defect detection. The performance of the network is measured by the MAP value. This proposed method yields the better MAP average 90 percent which is higher than the 36 percent and the 7 percent of the ZF and Resnet50 respectively. Future work may focus on defect detection in order to improve the accuracy rate as well as detection speed further.

References

1. Rinker Materials (2009) "Storm Sewer Pipeline Laser Profiling." Available at: www.rin kerpipe. com (accessed November 19, 2017).
2. ASCE "Infrastructure Report Card." Available at: www.asce.org/infrastructure/.
3. J. Mattsson, A. Hedström, M. Viklander, and G.T. Blecken (2014) "Fat, Oil, and Grease Accumulation in Sewer Systems: Comprehensive Survey of Experiences o Scandinavian Municipalities." *Journal of Environmental Engineering*, 140 (3). http://dx.doi.org/10.1061/(ASCE) EE.1943-7870.0000813.
4. C. Cortes and V. Vapnik, (1995) "Support-Vector Networks." *Machine Learning* 20(3): 273–297.: https://doi.org/10.1007/bf00994018.
5. EPA (n.d.) "Why Control Sanitary Sewer Overflows?" Available at: www3.epa.gov/npdes/pubs/ sso_casestudy_control.pdf.
6. Jack C.P. Cheng and Mingzhu Wang (2018) "Automated Detection of Sewer Pipe Defects in Closed-Circuit Television Images Using Deep Learning Techniques." *Automation in Construction*, 95: 155–171.
7. H. Zakeri, F.M. Nejad, and A. Fahimifar (2016) "Image Based Techniques for Crack Detection, Classification and Quantification in Asphalt Pavement: A Review." *Archives of Computational Methods in Engineering*, 24: 935–977. https://doi.org/10.1007/s11 831-016-9194-z.
8. G. Li, S. He, Y. Ju, and K. Du (2014) "Long-Distance Precision Inspection Method for Bridge Cracks with Image Processing," *Automation in Construction*, 41: 83–95. https://doi.org/10. 1016/j.autcon.2013.10.021.
9. A. Hawari, M. Alamin, F. Alkadour, M. Elmasry, and T. Zayed (2018) "Automated Defect Detection Tool for Closed Circuit Television (CCTV) Inspected Sewer Pipelines." *Automation in Construction*, 89: 99–109, https://doi.org/10.1016/j.aut con.2018.01.004.
10. R.J. King Wong Allen (2009) *Hong Kong Conduit Condition Evaluation Code*, 4th edn. Hong Kong: Hong Kong Institute of Utility Specialists.
11. D.H. Ballard and C.M. Brown (1982) *Computer Vision*. Englewood Cliffs, NJ: Prentice Hall.
12. R. Girshick (2015) "Fast R-CNN." Paper presented at IEEE International Conference on Computer Vision (ICCV), IEEE, Santiago, Chile, pp. 1440–1448, https://doi.org/10.1109/.
13. C. Koch, K. Georgieva, V. Kasireddy, B. Akinci, and P. Fieguth (2015) "A Review on Computer Vision Based Defect Detection and Condition Assessment of Concrete and

Asphalt Civil Infrastructure." *Advanced Engineering Informatics*, 29(2):196–210. https://doi.org/10.1016/ j.aei.2015.01.008.

14. R. Girshick (2015) "Fast R-CNN." Paper presented at IEEE International Conference on Computer Vision (ICCV), pp. 1440–1448. https://doi.org/10.1109/ ICCV.2015.169.

15. D.G. Lowe (2004) "Distinctive Image Features from Scale-Invariant Keypoints." *International Journal of Computer Vision*, 60(2): 91–110, https://doi.org/10.1023/ B:VISI.0000029664. 99615.94.

16. N. Dalal, and B. Triggs (2005) "Histograms of Oriented Gradients for Human Detection." Paper presented at IEEE Computer Society Conference on Computer Vision and Pattern Recognition (CVPR), IEEE, pp. 886–893, https://doi.org/10.1109/ CVPR.2005.177.

17. L. Zhu, Y. Chen, A. Yuille, W. Freeman (2010) "Latent hierarchical structural Learning for Object Detection." Paper presented at IEEE Conference on Computer Vision and Pattern Recognition (CVPR), IEEE, pp. 1062–1069, https://doi.org/10.1109/ CVPR.2010. 5540096.

18. P.F. Felzenszwalb, R.B. Girshick, D. McAllester, and D. Ramanan (2010) "Object Detection with Discriminatively Trained Part-Based Models." *IEEE Transactions Pattern Analysis and Machine Intelligence*, 32(9): 1627–1645. https://doi.org/10.1109/ TPAMI.2009.167.

19. F. Zana and J. Klein (1997) "Robust Segmentation of Vessels from Retinal Angiography." Paper presented at 13th International Conference on Digital Signal Processing, 1997. http://dx.doi.org/10.1109/icdsp.1997.628554.

20. R. Girshick, J. Donahue, T. Darrell, and J. Malik (2014) "Rich Feature Hierarchies for Accurate Object Detection and Semantic Segmentation." Paper presented at 2014 IEEE Conference on Computer Vision and Pattern Recognition, IEEE, Columbus, OH, USA, pp. 580–587, https://doi.org/10.1109/CVPR.2014.81.

21. S. Ren, K. He, R. Girshick, and J. Sun (2017) "Faster R-CNN: Towards Real-Time Object Detection with Region Proposal Networks." *IEEE Transactions on Pattern Analysis and Machine Intelligence*, 39(6): 91–99, https://doi.org/10.1109/TPAMI.2016.2577 031.

22. J. Myrans, R. Everson, and Z. Kapelan (2018) "Automated Detection of Faults in Sewers Using CCTV Image Sequences." *Automation in Construction*, 95: 64–71.

23. A. Hawaria, M. Alamina, F. Alkadoura, M. Elmasryb, and T. Zayed (2018) "Automated Defect Detection Tool for Closed Circuit Television (CCTV) Inspected Sewer Pipelines." *Automation in Construction*, 89: 99–109.

24. S. Kumara, Dulcy M. Abrahama, Mohammad R. Jahanshahia,T. Iseleyb, and J. Starrc (2018) "Automated Defect Classification in Sewer Closed Circuit Television Inspections Using Deep Convolutional Neural Networks," *Automation in Construction*, 91: 273–283.

Chapter 13

Learning and Reasoning Using Artificial Intelligence

Merjulah Roby

13.1 Introduction

Artificial Intelligence (AI) is based on the workings of the human brain, in order to build smart machines. AI is capable of performing work intelligently without human intervention, and is becoming capable of acting and thinking like a human. AI involves natural language processing, speech recognition, expert systems and machine vision to behave like a human. Human intelligence involves the mental capability which includes planning, the skill to learn, to solve difficulties, to understand complex ideas, to be quick at learning, to be able to reason abstractly and to learn through practice. AI is not learning from a book or academic skills. Rather, it replicates a larger and deeper understanding of its surroundings.

13.1.1 Human Intelligence Based on the Psychological View

Human intelligence is the mental quality which includes the ability to learn through experience, adapt to new situations, understand concepts and handle them and finally use the knowledge to manipulate the environment. Different investigators have offered different features of intelligence in their own definition. In 1921, American psychologists, Lewis Terman and Edward L. Thorndike, had different opinions on the definition of intelligence. Terman strongly believed in the ability to

DOI: 10.1201/9781003424550-13

think, while Thorndike emphasized learning and the ability to offer good responses to questions. In general, the psychologists agreed that adapting to the environment is the key understanding of intelligence. This adaption has a number of intellectual processes such as perception, learning, memory, reasoning and solving problems. Thus, intelligence is not a single ability but the combination of many abilities. AI uses computers to simulate human intelligence.

13.1.2 Types of Artificial Intelligence Based on Functionality

Artificial Intelligence is mainly of two types based on functionality. Type 1 is categorized into narrow AI, general AI and strong AI. Figure 13.1 shows the types of AI based on functionality.

13.1.2.1 Type 1 Category

AI includes the ability to think like humans, solve puzzles, to reason, plan, make judgments, learn and communicate on its own.

■ *Narrow AI*: Narrow AI is one of the AI types which performs a task based on intelligence. The most commonly used AI is narrow AI. It is trained only for a specific task, so it cannot perform beyond the trained task. Hence it is called weak AI. It fails in unpredictable ways when the task is beyond the limit of

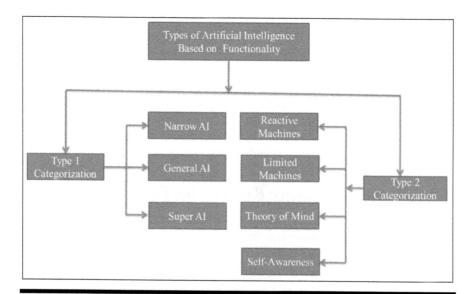

Figure 13.1 Types of AI based on functionality.

training. Watson is an IBM supercomputer that is combined with narrow AI, which uses Natural Language Processing (NLP) and Machine Learning (ML). Examples of narrow AI are e-commerce sites for purchasing, image recognition, speech recognition, autonomous car self-driving, or playing chess, etc.

■ *General AI*: General AI is a type of intelligence which can perform any kind of task with efficiency like a human. The general idea behind the AI is to act smart and think like a human on its own. Researchers are focusing on the development of the machine with General AI. It is still at the construction stage and it will take much time and effort to develop such a system.

Super AI; Super AI is the level of intelligence which could exceed human intelligence and could perform a task better than a human, along with possessing cognitive properties. It is the result of General AI development. Super AI is still a theoretical concept of AI. Development of Super AI is still a challenging task in the real world. The second main type of AI is the Type 2 category which is divided into reactive machines, limited machines, theory machines and self-aware machines.

13.1.2.2 Type 2 Category

■ *Reactive machines*: Reactive machines (RM) are the basic types of AI. This AI system will not store memories for future use. This system focuses on the current scenario and performs the best action. An example of an RM is the Deep Blue System by IBM and AlphaGo by Google.

■ *Limited machines*: Limited machines (LM) store the experience and the data for a short period of time. An example of an LM is the self-driving car. These cars store the recent driven speed and the navigation information.

■ *Theory of mind*: Theory of mind AI should recognize human emotions, beliefs and people and communicate like humans. These kinds of system are still under development.

■ *Self-awareness*: Self-awareness is the future of AI. This machine will be super-intelligent and have sentiments, consciousness and even self-awareness. These machines are smarter than humans.

13.1.3 Types of AI Based on Technology

There are three types of AI based on technology: Artificial Narrow Intelligence (ANI), Artificial General Intelligence (AGI), and Artificial Super Intelligence (ASI). Figure 13.2 shows the types of AI based on the technology.

■ *Artificial Narrow Intelligence* (ANI) is the most used AI. ANI performs one or two tasks based on the training data and its learning experience through the previous tasks. Currently, the core development of AI used for decision-making

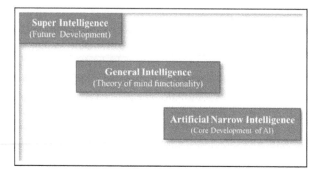

Figure 13.2 Types of AI based on technology.

can be categorized under ANI. It is also called weak AI as it completes the task under a predefined and limited set of constraints.

- *Artificial General Intelligence* (AGI) is related to the theory of mind functionality. The AGI is still in the development stage. The main target of AGI is to create independent connections through various domains.
- *Artificial Super Intelligence* (ASI) is the future development of AI achievement. AI will improve in memory performance and better decision-making. It will provide an outstanding amount of memory, a fast processing capacity and enhanced way of intelligent decision-making.

13.1.4 Types of Intelligence

The intelligence system is a branch of AI which describes the calculation ability, reasoning, perception, the ability to store and retrieve the data through memory, complex problem solving and adaption.

- *Linguistic intelligence*: This is the skill to speak, recognize sounds and the mechanism of phonology such as speech and sound, grammar syntax and meaning. Narrators and Orators are examples of linguistic intelligence.
- *Musical intelligence*: The skill to create, communicate and also understand the meaning of sound, pitch and rhythm. Musicians, singers and composers are examples of musical intelligence.
- *Logical mathematical intelligence*: This is the skill to use and understand the relationship in the absence of the action or objects and understand complex ideas. Mathematicians and scientists are examples of logical mathematical intelligence.
- *Spatial intelligence*: This is the skill to perceive visual or spatial information, recreate visual images without reference to the objects, use 3D image

construction and to rotate and move them. Astronauts, map readers and physicians are examples of spatial intelligence.

■ *Bodily-kinesthetic intelligence*: This is the skill to solve the complete or part of body problems or products of fashion, the ability to control motor skills and object manipulation. Sports players and dancers are examples of bodily-kinesthetic intelligence.

■ *Intra-personal intelligence*: This is the skill to distinguish among one's own feelings, purposes and inspirations. Gautam Buddha is an example of intra-personal intelligence.

■ *Inter-personal intelligence*: This is the skill to recognize the variation in other people's feelings, purposes and beliefs. Mass communication and interviewers are examples of inter-personal intelligence.

13.1.5 Structure of Intelligent Agents

The intelligent agent is the autonomy to complete specific, repetitive and predictable tasks for users and applications. It is also called intelligent because of its learning ability based on experience. The two main functions of the structural intelligent agents (SIA) are action and perception. Actions are performed through actuators and the perception is through the sensors. Intelligent agents also have sub-agents which forms a hierarchical structure. The lower tasks are performed by the sub-agent. Both the higher- and the lower-level agents combined form a system which could solve multiple problems through the intelligent response and behavior.

The features of intelligent agents are:

■ Level of autonomy helps to perform certain work.
■ Learning ability through prior experience.
■ Interaction with agents, humans and systems.
■ Intelligent agents can adopt to the new rules incrementally.
■ Goal-oriented system.
■ Intelligent agents are knowledge-based through processing, communication and entities.

The intelligent agents have five categories. Grouping of these agents is based on their capabilities and the level of superficial intelligence. Figure 13.3 shows the categories of intelligent agents. The five categories are:

1. Simple reflex agent.
2. Model-based reflex agent.
3. Goal-based agent.
4. Utility-based agent.
5. Learning agent.

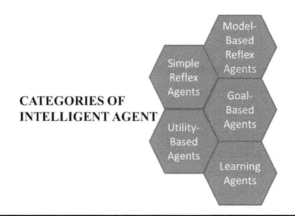

Figure 13.3 Five categories of the intelligent agent.

1. *Simple reflex agents*: The simple reflex agents use the current experience rather than the past performed history. The condition action rule is the basis for the function of an agent. The success of the agent function is because the observational environment is ideal.
2. *Model-Based Reflex Agents (MRA)*: The MRA considers the historical data in its actions. The MRA can perform well even if the environment is not fully observable. This agent uses the internal model which determines the history and the effects of action. It reflects on the current state that was unobserved.
3. *Goal-based agents* have higher capabilities than the previous agents. They use the goal information for the capability description. These allow it to choose between various possibilities. These agents select the best possibility which enhances the performance of the goal.
4. *Utility-Based Agents* make decisions based on utility. These agents are more advanced than the goal-based agents. Using a utility function, a state is mapped against the measure of utility. A rational agent selects the action which optimizes the utility of the outcome.
5. The *learning agents* have the capability to learn through the previous experience. The elements of learning agents are as follows:
   a. Learning element: This element makes the agents learn through previous experiences.
   b. Critical element: It provides feedback on the agents' performance.
   c. *Performance element*: This depends on the external action which needs to be taken.
   d. *Problem generator*: It provides the feedback agent which performs some tasks, such as making suggestions and keeping track of the history.

13.1.6 How Intelligent Agents Work

Intelligent agents work through three components: sensors, actuators and effectors. An overview of these components can give a clear idea of how the intelligent agents work.

- *Sensors:* Sensors detect the changes in the environment. The details are sent through other devices. In AI the environment is observed through the sensor agent.
- *Actuators* – Actuators convert the energy into motion and perform the role of moving and controlling a system. Examples are motors, rails and gears.
- *Effectors:* The effectors affect the environment. Examples are fingers, legs, wheels, display screens and arms.

Figure 13.4 shows the inputs from the environment are received through the intelligent agent by sensors. The sensor agent uses AI to make decisions based on the observation through experience of prior tasks. Then action is activated through actuators. Future decisions are based on history and past actions.

13.1.7 Intelligent Agent Applications

Intelligent agents in AI are applied in many real-life situations:

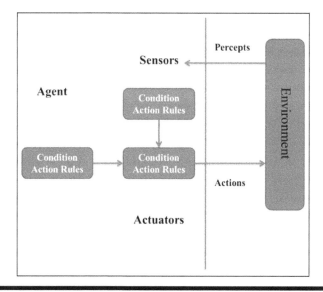

Figure 13.4   Components' position in AI.

■ *Searching for information, retrieval and navigation of data*: Intelligent agents enrich the navigation and access the information. This is done through the search engine. Internet objects take more time for the search of specific data objects but the intelligent agents perform this task in a very short time.

■ *Repetitive office activities*: Many companies have automated the administrative tasks to reduce costs. Examples of automated areas are customer support and sales. Intelligent agents have also enhanced the office productivity.

■ *Medical diagnosis*: Intelligent objects have also been included in the medical domain (healthcare services) to improve the patients' health. In this field the patients are the environments. The keyboard of the computer is the sensor which receives the patient's symptoms. The intelligent agent decides on the best course of action. Medical care is provided through the actuators, such as tests and treatments.

■ *Vacuum cleaning*; Agents in AI are also used to vacuum areas with increased efficiency. In this field, a room, table or carpet is the environment. Cameras bump sensors and dirt sensors are some of the sensors. Wheels, brushing and vacuum extractors are the actions by actuators.

■ *Autonomous driving*: Intelligent agents improve the functions of self-driving cars. In the automatic driving system, sensors are activated to gather information through the environment. They include GPS, cameras and radar. In this field the pedestrians, roads, vehicles and road signals are detected. To initiate actions various actuators are used. Example, the brake in the car to stop the car.

13.1.8 Artificial Intelligence in Everyday Application

AI has been developing over the past decades and it is enabling businesses and people to think beyond their thinking. Equipping the machine to think and evaluate is one of the challenges in the computer world. The adoption of AI and machine learning has eliminated many repetitive tasks and turned the focus on the critical problem. AI is the broad term used to summarize a variety of subdomains, and machine learning is one of them. AI and machine learning are used widely in businesses for solutions. Online recommendation and online searches are applications of AI. Other applications of AI cover a wide range of industries and other areas:

■ *Virtual assistant*: The virtual assistants, like Amazon's Alexa, Google's assistant and Apple's Siri, use the voice-to-text technology and AI to perform searches, with connectivity based on the internet, for online orders, list creation, reminder setting and answering questions. If integrated in the smart home and entertainment system, it can be used to automate the daily routine. Algorithms are used for the conversational interfaces to train the system to cater for the customer. Advanced chatbots could give an answer to the complex questions and give the customer the impression that they are talking with a customer representative. AI in web applications reduces the labor costs which saves money for the commercial organization.

■ *Farming and agriculture*: Crop production is increased through AI monitoring, based on the growth of the plant and weather monitoring. In the agricultural sectors, robots, autonomous tractors, drones are helping to monitor the health condition of the crop and also implement harvesting, productivity enhancement and cultivate farmland.

■ *Automatic self-driving*: Automatic self-driving in the car is connected to the sensors and gathers the information through Google Street View and the cameras inside the car. The AI simulates the human interpretation and decision-making processes based on deep learning technique and controls the actions in the driving system. Self-driving developers use large amounts of data through image recognition combined with the neural network and machine learning network to build the framework to drive automatically. The neural network identifies the data patterns and makes the machine learn the patterns. The data of the cameras on the car includes trees, traffic lights, curbs, street signs, pedestrians, and the other environmental surroundings on the road during driving. To qualify as fully automatic the self-driving vehicle must drive without the help of the human. Companies developing autonomous cars are Audi, BMW, General Motors, Volkswagen, Tesla and Volvo. The self-driving features of an autonomous car are:

■ Hands-free steering: This allows the car to drive without the driver's hand on the steering wheel. But the attention of the driver is required.

■ Cruise control: This automatically maintains the distance between the driver's car and the car in front.

■ Auto steer: This is an advanced driver assistant system which keeps the road vehicle centered in the lane.

13.2 Growing Use of AI in Online Applications

13.2.1 E-Commerce

AI in online shopping is a big advantage in the field of e-commerce. The chatbot services, customer comments analyzer and also the personal services for the online shopping are provided by e-commerce. For example, if a person repeatedly orders the same brand of rice, then the online retailer could send a personalized offer to the shopper for that product or use the machine learning technique to recommend supplementary products. Shopping online is a very comfortable and easy way for people to shop using the internet from their home. Different manufacturers put their products on the website. Customers browse the website, place the order and make the payment using different options like credit card or debit card payment.

13.2.2 E-Learning Tools

The digital era enhances different areas in society, also including education. Online education has developed in the past few years, based on different types of interactive tutorials, e-books, video tutorials, etc. Moreover, even one-to-one interaction with students makes the platform more comfortable and also offers individual attention and understanding of the subject.

13.2.3 Conducting Auctions

Numerous websites provide products for auction. Customers bid for the product and make the payment according to their bidding amount. E-bay is one such website which provides the bidding option for the product.

13.2.4 Travel Reservations

Booking tickets for flights or hotels is one of the popular uses of e-commerce. Online travel reservation is one of the most convenient ways to save time and to find the different options for destinations and offers.

13.2.5 Online Stock Trading

Online trading is where business uses the internet to buy and sell stocks and shares. The brokers trade electronically on the stocks and in real time. The computer screens are links, where the brokers connect the buyers and the sellers. This is one of the methods to reduce costs.

13.2.6 Electronic Banking

The mobile app for electronic banking has been introduced by many banks. Using the mobile or through the computer, customers can connect to the bank and control their financial dealings in the comfort of their own surroundings. They can even pay many bills through these facilities. This kind of service reduces the work of the staff.

13.2.7 Advertising and Marketing

E-commerce provides an important platform for advertising and marketing new products. E-commerce uses popular websites to project advertisements for the new product launch to make the customers aware of new products.

13.2.8 Customer Service

Businessmen get an opportunity to communicate with the customers through the internet. Through online messages, the product defects can be identified through the reviews by the customers. Complaints can be rectified through various services based on the customers' feedback.

13.3 AI and Security and Surveillance

CCTV cameras for security and surveillance installation are very common now. Cameras are considered the fundamental commodity to set up the surveillance infrastructure, but monitoring for 24/7 thousands of videos by operators doesn't assist the purpose of providing proactive surveillance and fast response to events.

The real-time warnings for standard problems like detection of motion, left baggage, etc., are raised based on the software-based video content analytics. The inaccuracy and the false positives when using video content analytics have made the operators disable them to avoid false alarms.

AI and deep neural network (DNN), along with the video content analytics, are trained to identify, detect and differentiate various obstacles in videos with the help of large tagged examples. In addition to extracting the data speed and size, direction, path, color and area, the AI-based object classification uses computer vision algorithms. The data can be searched to concentrate the video efforts on analytics on appropriate information.

In the last decade, in the field of computer vision, the experts have practically implemented the deep learning of theoretical ideas with the amount of data and increased computational power.

13.4 AI and Medical Image Processing

AI supports the radiologist and the pathologist to analyze medical imaging. AI and machine learning have captivated the health industry and produce accurate results and are used to perform many tasks. AI is helping to uncover hidden data for clinical decision-making. Medical imaging is a rich source of information about the patients with complex conditions. The pixels of information packed into the results of an MRI, X-rays, PET or CT scans and other imaging modalities, combined with high resolution images, can be of great assistance to the medical experts.

13.4.1 Cardiovascular Abnormalities

In recent years, using the AI technique, cardiac imaging analysis has greatly developed. AI can help to analyze the results of echocardiography, coronary angiography, and electrocardiogram. The different size of the heart shows the cardiovascular risk in each individual. Automating the identification of the abnormality in common imaging tests including chest X-rays, MRI, etc. could lead to quicker decision-making with few errors. In addition, AI can also analyze echocardiographic images that include the measurement of the size of each chamber and the assessment of functions of the left ventricle.

AI can also help the physician to make a quick decision on the identification of the sub-clinical organ dysfunction with the help of the clinical-related information. The continuous development of AI with the sub-domains machine learning and deep learning makes it attractive to medical experts as it can create a reliable and efficient model for healthcare.

Echocardiography is critical in diagnosis and management of cardiovascular diseases and the accurate assessment of cardiac functions and structures. AI tools provide new features in machine learning to enhance accuracy, based on image interpretation. Machine learning helps in automatic interpretation of the unused data generated by the multidimensional modality of imaging.

A wide range of machine learning models is used for the diagnosis of cardiovascular disease in clinical echocardiography practice. Sengupta et al. [1] proposed a cognitive machine learning technique with the speckle tracking echocardiographic data which differentiates constrictive pericarditis from restrictive cardiomyopathy. The research work on the interpretation of the speckle tracking echocardiographic data is continuing. Naurla et al. [2] developed a supervised learning algorithm to differentiate the heart of athletes and hypertrophic cardiomyopathy based on the speckle tracking echocardiographic data. Another application of heart valve disease using machine learning model is given by [3, 4]. Playford et al. [5] estimated whether AI could analyze the aortic valve area in aortic valve stenosis without measuring the left ventricular outflow track from other echocardiographic data with an accuracy of 95 percent.

In cardiac MRI, segmentation of ventricles in the heart is one of the fields which are supported by machine learning models. AI improves the efficiency of the clinical assessment [6–8]. Avendi et al. [9] proposed a deep learning technique for automatic detection and segmentation of the right ventricular chamber, based on the cardiac MRI dataset. In the same way, various automated neural network techniques are used for the segmentation of the right ventricle based on the cardiac cine MRI [10–11]. Dawes et al. [12] proposed the supervised machine learning of systolic cardiac motion for an early prediction of right heart failure in patients with pulmonary diseases.

Machine learning algorithms in CT are increasingly used in the diagnosis of risk assessment of coronary artery disease and in atherosclerosis. Coronary computer tomographic angiography is a non-invasive medical modality to detect coronary artery disease. Various machine learning models have been developed [13] to control non-invasive fractional flow reserve and increase the performance of coronary computer tomographic angiography. Wolterink et al. [14] used supervised machine learning to identify coronary artery calcification. Machine learning analysis is performed on prospective and retrospective data, such as clinical, coronary computer tomographic angiography imaging, biohumoral, lipidomics, etc. to distinguish the high risk patients.

13.4.2 Musculoskeletal Injury

Imaging is one of the important tools for the identification of patients with musculoskeletal conditions. Increased usage of imaging includes an operational efficiency with better accuracy and better quality of the image. AI is an exciting tool to help the radiologists meet their requirements.

13.4.3 Neurological Diseases

The neuro-oncology field has many challenges and AI can help to overcome them. AI helps in the field of neurodegeneration that includes Alzheimer's disease, Parkinson's disease, or amyotrophic lateral sclerosis. AI supports in the assessment of brain tumors and their diagnosis. Heidelberg University and the German Cancer Research Centre trained the machine learning algorithm, based on the 500 MRI scans of patients with brain tumors. Volumetric tumor segmentation is the result, due to the algorithm used to detect brain tumors automatically based on the MRI scan. The AI technique has great value, offers an accurate diagnosis, which helps in finding the response of the therapy. Prediction is one of the other applications in neuro-oncology. AI is used to speed up the process based on brain mapping or functional brain imaging.

13.4.4 Thoracic Condition and Complications

In this time of rapid development of AI, surgeons and researchers thought AI could contribute to healthcare in all aspects and especially in surgery. The advanced method of AI-assisted technology facilitates surgical teaching, simulation and also planning. A new surgical application based on surgical videos is assisted through AI, which is still in the developmental stage. AI has the potential application in education in thoracic surgery, with evaluation based on surgical quality.

There are various common human primary malignant tumors in thoracic diseases, among which are esophageal cancer and lung cancer, which are the serious cancer events and are fatal. Early analysis is important for cancer treatment, so AI systems have been developed for automated and accurate detection and the diagnosis of thoracic tumors.

13.5 Advantages of AI

13.5.1 Automation

Automation is one of the source benefits of AI models which has an influence on the transportation industry, communications, the service industry, and consumer products. Automation leads to high accuracy along with the efficient use of the raw materials, to improve production quality, lead time reduction and offer superior safety.

13.5.2 Smart Decisions by AI

AI makes smarter decisions for businesses. AI includes data delivery, analysis of the trends, developing data consistency, providing forecasts, and assessing uncertainty when making a better decision. If AI is not supported to imitate the human emotions, it remains unbiased on the matter, which helps to make better decisions for the development of the system.

13.5.3 Increased Customer Experience

AI can help businesses respond to customer queries and it addresses the situation quickly. The chatbot with natural language processing helps to communicate and can generally give a personalized message to the customers who are searching for the best solutions to their problems. AI also reduces the stress of the customers through a customer support system which increases productivity.

13.5.4 Medical Support System

The use of AI in the medical field is increasingly popular these days. Remote analysis of the patient by monitoring technology allows the healthcare sector to perform diagnosis and quick treatment without the actual presence of the patients. AI also can be useful in monitoring communicable diseases and predict the results of the future outcome.

13.5.5 Data Analysis

AI and machine learning are used to analyze the data in an efficient way. It helps in creating the model to interpret and process the data for the future outcome. The advanced capability of the computer can speed up the process as the manual interpretation by humans takes too long to understand the data. The AI and machine learning techniques are designed to solve complex problems. These include interaction with customers, detection of fraud, weather reporting and medical diagnosis. Also, AI helps in finding the best solutions to the challenges which increases productivity and reduces costs.

13.5.6 Stability of Business

Business forecasting based on AI helps to make critical decisions, even in emergencies, to ensure the stability of the business. In general, there are many benefits for business by using AI to increase efficiency. AI has a great ability to gather vast amounts of data and automatically does all the work, based on the training and implementation. AI can increase the business performance in the following ways:

- sales prediction
- extracting and reviewing data
- leveraging smart chatbots
- AI tools implementation
- automatic call management
- real-time operations.

13.5.7 Managing Repetitive Tasks

Businesses are using AI to increase the productivity of their employees. One of the major benefits of AI for businesses is handling repetitive processes in the organization, which frees the employees to focus on creative and complex solutions. One of the best examples of the repetitive processes is the chatbots. An AI-operated chatbot helps in daily interactions with customers and offers fast and accurate support to the customers.

13.5.8 Minimizing Errors

AI solves complex problems consistently and also speeds up the process which is unmatched by human intelligence. But AI can't replace human intelligence, it needs the right problem to solve, the right data is required, in order to design the solutions for the problem and develop the right process for AI to adapt, to learn through errors and produce better results. AI also provides more time for the employees to handle the impactful task of the business. Even the wrong diagnosis of the doctors can be reduced, based on the large amount of data which connects the symptoms of illness and the cause. AI strengthens the decision of the doctors and removes the error, based on machine learning and the pattern recognition algorithm.

Undoubtedly, healthcare and other businesses around the world are leveraging AI to produce important leaps in error management.

13.6 Training AI

When we train AI, we are teaching the model to interpret the data and learn through it to perform the task with accuracy. Only through training to correctly interpret the data will AI make accurate decisions based on the information provided. All machine learning projects require high quality data in order to be successful. If the AI is trained on poor features, then the accuracy of the output will be poor quality of AI. AI training has three stages. Figure 13.5 shows the three stages of the AI model: the training stage, the validation stage and the testing stage.

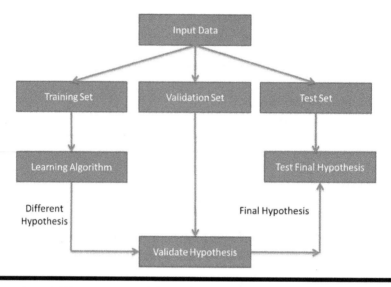

Figure 13.5   Different stages of the AI framework.

1. *Training stage*: In the first training stage, an AI model is given a set of training data and makes decisions based on the trained information. AI sometime stumbles in this stage. Like a child, it is beginning its learning. As we point out the mistakes, the adjustment has to be made which helps AI become more accurate. One of the main things to avoid is overfitting. Overfitting happens when a model learns the noise details in the training stage that negatively affect the performance of the new data.

2. *Validation stage*: Once the AI has completed the training stage, the next stage is validation. This stage provides the information of how the AI performs using the new set of data. As with the training stage, one has to evaluate and confirm the behavior of AI and the new variable which is not considered earlier. Issues with the overfitting will be revealed during the validation stage.

3. *Testing stage*; After the training stage, the new dataset is sent to the testing stage to see the performance of the AI model. The raw dataset is sent to the AI model for testing, to see if AI can make perfect decisions on the unstructured information. If the decision is wrong, the training stage has to be repeated until it is performed accurately.

13.6.1 Success of AI Training

In general, we need three key features to train the AI model to work well: high quality data, accurate annotation of data, and experimentation.

1. *High quality data*: The AI model requires high quality data. If the dataset includes a small amount of poor quality data, then there is a chance it will produce undesirable results. The situation of undesirable results yields error in AI called bias.
2. *Accurate annotation of data*: The accurate interpretation of the data is more essential than having high quality data. Otherwise, the AI model learns the wrong interpretation. Correctly annotated images help AI to learn better and perform with high accuracy.
3. *Experimentation*; Expect the AI to make mistakes during the training stage. Errors also called bias are normal during the AI training process. When AI does not interpret the data properly, the main key is to analyze the result which helps to identify what went wrong during the training stage. The knowledge received through deep identification help to improve AI, which leads to better prediction with high accuracy.

13.6.2 Train, Test and Maintain AI and Machine Learning Models

The model creation process is required for the skill sets of creating the AI and ML, which is performed by the ML software. The predefined success criteria compose the challenge to produce the model.

ML software requires the data to train the model, with the model establishing the AI product which can be re-used over time, based on updates of the AI input. The ML software has four simple learning types:

1. *Supervised learning*: Making the algorithm learn the data based on the labels presented on the data. This in general means that the values for the prediction are known, which are well defined from the beginning for the algorithm.
2. *Unsupervised learning*: Unlike the supervised methods, the algorithm doesn't have any labels or have the correct answers for the data; it is based on the algorithm's options which bring together the related data and understands it.
3. *Semi-supervised learning*: A hybrid learning of supervised and unsupervised learning.
4. *Reinforcement*: Rewards are given to the algorithm for the correct prediction to drive the accuracy high in reinforcement learning.

Data scientists are needed to determine the best statistical algorithms when using the ML software. The most popular statistical algorithms are Naïve Bayes which is used for sentimental analysis, detection of spam and for recommendations. Decision tree is used for outcome predictions to improve the prediction performance, the random forest algorithm is used which merges multiple decision trees. The logistic regression is for binary classification. The linear regression is used to categorize a large dataset.

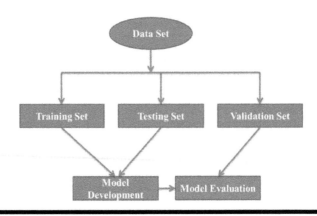

Figure 13.6 Model building.

Clustering the organized data into groups like market segmentation uses Gaussian mixture, K-Means Clustering AdaBoost, and Recommender. Figure 13.6 shows the model building of the data set.

There are three stages of learning based on the machine learning: training, validation and testing. Before it starts, it's necessary to safeguard the data, making it well organized and perfect. Getting the data transformed into order is time-consuming and requires human intervention for the detailed-oriented process. The main goal is for thee data to be free of data duplication, disconnection and typos. The data is divided into three sets after cleaning which is used for the three stages of training. Random data division is to discourage the selection of data biases.

Here are a few definitions relevant to the creation of the model:

- *Parameter:* Parameters are automatically learned values by the ML software through the input data and also the parameters are manually changeable during the training process by the user. Examples are the maximum number of passes made during the session and the maximum model size of the data for training is in bytes.
- *Hyperparameter:* The external value of the ML is the hyperparameter. The data scientist as a user gives the input before the process, so the hyperparameter values are not processed from AI data which are changeable during the training process. Examples of hyperparameters are the number of layers in the neural network, the clustering algorithm, the number of clusters to be returned.
- *Variable:* The variable is the data input fields chosen as the value by the ML software, with potentially using variables additionally as the training progresses. Examples of variables are age, weight and height.

Before the training starts, the initial stage is to label the data so the ML software can process it with the help of vital clues. Unsupervised learning doesn't require the labels which are added later. Default parameter values of ML software can start the process or the parameters can be modified individually.

It is in the validation or the training stage that AI encounters the success criteria. The new set of data is used in the first pass, if the results performed better, then it proceeds to the final stage, which is testing. If the results performance is low, then the ML software makes additional passes through the data, continues till the ML software displays no patterns or it reaches the maximum number of passes. The parameters are modified automatically by the ML software or managed by advancing the training.

The final stage is the testing stage of a new set of data for supervised learning. It will become a working model when the software passes the success criteria test. If it fails, it has to go back to the training stage. The team will manually change the parameters or the ML software modifies the parameters automatically at the training stage.

A repetitive play of the ML software on the dataset is AI machine learning, automatically changing parameters modifies the data iteratively by the ML software and by human interventions to make the model smarter after each data pass. Multiple passes are the work of the ML software until no new patterns are detected or it reaches the maximum number of passes which causes it to stop.

AI freedom is the constant monitoring. To determine the performance of the AI model requires the obvious task of monitoring which leads to actual performance matching the AI prediction. If the prediction performance worsens, it has to re-enter the ML training model process for the correction of the up-to-date data.

The input data are easily changeable over time, which is called data drift. Data drift causes the accuracy of the AI model to deteriorate, data drift warnings should be identified ahead of the problems. The tools of AI can track the data drift which finds the outlier. Neptune, Fiddler and Azure ML can provide early warnings so the data problems can be addressed by updates in the ML sooner rather than later.

13.7 Conclusion

We are on the cusp of a revolution of the artificial intelligence and there is a lot more to come in future. There are many research studies on AI, and a detailed history would include AI applications in public service, military applications based on AI, the ethics of AI and the rules of robotics. AI is increasingly integrated into our daily lives from video games, computers and kitchen appliances. However, there are problems and challenges with AI. There are pros and cons of every new technology and AI is also no exception. Therefore, it is concluded that AI is a great technology which helps in new data science, engineering, machine learning and IT. It is positioned for the development and maintenance of system software which runs on AI algorithms and enhances the quality of human life.

References

1. Partho P. Sengupta, Yen-Min Huang, Manish Bansal, Ali Ashrafi, Matt Fisher, Khader Shameer, Walt Gall and Joel T. Dudley (2016) "Cognitive Machine-Learning Algorithm for Cardiac Imaging: A Pilot Study for Differentiating Constrictive Pericarditis from Restrictive Cardiomyopathy." *Circulation: Cardiovascular Imaging*, 9(6).

2. Sukrit Narula, Khader Shameer, Mabrouk Salem Omar, Joel T. Dudley, and Partho P. Sengupta (2016) "Machine-Learning Algorithms to Automate Morphological and Functional Assessments in 2D Echocardiography." *Journal of the American College of Cardiology*, 68(21): 2287–2295.

3. R. Sengur (2012) "Support Vector Machine Ensembles for Intelligent Diagnosis of Valvular Heart Disease." *Journal of Medical Systems*, 36(4): 2649–2655.

4. C. Martin-Isla, V. M. Campello, C. Izquierdo et al. (2020) "Image-Based Cardiac Diagnosis with Machine Learning: A Review." *Frontiers in Cardiovascular Medicine*, 7.

5. E. Playford, L. Bordin, R. Talbot, B. Mohamad, B. Anderson, and G. Strange (2018) "Analysis of Aortic Stenosis Using Artificial Intelligence." *Heart, Lung and Circulation*, 27: S216.

6. H. B. Winther, C. Hundt, B. Schmidt et al. (2018) "v-Net: Deep Learning for Generalized Biventricular Mass and Function Parameters Using Multicenter Cardiac MRI Data." *JACC: Cardiovascular Imaging*, 11(7): 1036–1038.

7. M. R. Avendi, A. Kheradvar, and H. Jafarkhani (2017) "Automatic Segmentation of the Right Ventricle from Cardiac MRI Using a Learning-Based Approach." *Magnetic Resonance in Medicine*, 78(6): 2439–2448.

8. L. K. Tan, R. A. McLaughlin, E. Lim, Y. F. Abdul Aziz, and Y. M. Liew (2018) "Fully Automated Segmentation of the Left Ventricle in Cine Cardiac MRI Using Neural Network Regression." *Journal of Magnetic Resonance Imaging*, 48(1): 140–152.

9. M. R. Avendi, A. Kheradvar, and H. Jafarkhani (2017) "Automatic Segmentation of the Right Ventricle from Cardiac MRI Using a Learning-Based Approach." *Magnetic Resonance in Medicine*, 78(6): 2439–2448.

10. Q. Tao, W. Yan, Y. Wang et al. (2019) "Deep Learning-Based Method for Fully Automatic Quantification of Left Ventricle Function from Cine MR Images: A Multivendor, Multicenter Study." *Radiology*, 290(1): 81–88.

11. T. A. Ngo, Z. Lu, and G. Carneiro (2017) "Combining Deep Learning and Level Set for the Automated Segmentation of the Left Ventricle of the Heart from Cardiac Cine Magnetic Resonance." *Medical Image Analysis*, 35: 159–171.

12. T. J. W. Dawes, A. de Marvao, W. Shi et al. (2017) "Machine Learning of Three-Dimensional Right Ventricular Motion Enables Outcome Prediction in Pulmonary Hypertension: A Cardiac MR Imaging Study." *Radiology*, 283(2): 381–390.

13. Y. Shiono, H. Matsuo, T. Kawasaki, et al. (2019) "Clinical Impact of Coronary Computed Tomography Angiography-Derived Fractional Flow Reserve on Japanese Population in the Advance Registry." *Circulation Journal*, 83(6):1293–1301.

14. J. M. Wolterink, T. Leiner, B. D. de Vos, R. W. van Hamersvelt, M. A. Viergever, and I. Isgum (2016) "Automatic Coronary Artery Calcium Scoring in Cardiac CT Angiography Using Paired Convolutional Neural Networks." *Medical Image Analysis*, 34: 123–136.

Chapter 14

A Novel Auto Encoder-Network-Based Ensemble Technique for Sentiment Analysis Using Tweets on COVID-19 Data

R. Jyothsna, V. Rohini, and Joy Paulose

14.1 Introduction

Rapid developments in technology have resulted in people being able to express their views and opinions freely across the social media networking sites (Daradkeh, 2022). Social media sites are a great platform for expressing emotive issues. Emotion Artificial Intelligence, also termed opinion mining or sentiment analysis, is a procedure that involves analyzing whether a piece of text conveys a positive, negative or neutral opinion regarding any subject matter under consideration (Paliwal et al., 2022; Shukla and Garg, 2022; Wu et al., 2022). The sentiments behind such emotive issues can be analyzed using several sentiment analysis methods like machine learning, lexicon-based, deep learning, hybrid and nature inspired techniques (Alali et al., 2022; Dang et al., 2021; Ruz et al., 2022). Among supervised and unsupervised machine learning techniques, unsupervised machine learning techniques are usually applied for sentiment analysis as they do not require training datasets. Clustering, an unsupervised machine learning technique, is quite popular (Kaushik and Bhatia, 202; Pandey et al., 2022). Several NLP methods can be adopted for evaluating the emotions considering any scenario (Dandekar and Narawade, 2022). Opinion mining plays a great role in generating several models that mimic the human thinking processes while handling uncertain situations (Kaur et al., 2022). Sentiment analysis is an indispensable part of cognitive computing (Pasupa and Ayutthaya, 2022). Opinion mining is multi-disciplinary in nature that involves the culmination of Natural Language Processing, Machine Learning, Deep Learning, Statistics, Mathematics, Psychology, Sociology, Artificial Intelligence and many other domains. Innumerable research articles are available on sentiment analysis. Several tertiary studies are also carried out using secondary research articles n opinion mining. The information obtained after implementing opinion mining on a certain subject matter can be incorporated to explain, analyze and comprehend that particular phenomena (Pozzi et al., 2017). Considering the business area, opinion mining plays a very important role in analyzing the customers' views on a particular product. Strategies to improve the business can be analyzed using sentiment analysis. Analyzing the needs of the customer is extremely important in the current business domain which is highly aligned towards the customers (Chagas et al., 2018).

14.2 Background and Related Work

The present work considers "sentiment analysis" to be a common terminology for sentiment analysis as well as opinion mining. Sentiment analysis revolves around the concept of Natural Language Processing (NLP) as suggested by Cambria et al. (2017); Hemmatian and Sohrabi (2017); and Pozzi et al. (2017). A three-tier architecture can be suggested considering sentiment analysis as follows:

1. *Syntactic layer.* The syntactic layer involves the following functions: normalization of the dataset under consideration, disambiguation of boundaries associated with the sentences, tagging parts of speech to each set of words in the sentence or phrase, breaking down an entire sentence into smaller chunks and lemmatizing the words. i.e., reducing each word to its root form (Hemmatian and Sohrabi, 2017).
2. *Semantics layer.* The semantics layer involves the following tasks: detection of subjectivity and objectivity of sentences, extraction of underlying concepts, anaphora resolution and recognition of named entities (Kumar and Teeja, 2012).
3. *Pragmatics layer.* The pragmatics layer involves the following functions: detection of sarcastic elements in the dataset, understanding the metaphors involved, extraction of aspects involved in a particular sentence and detection of polarities (Cambria et al., 2017; Ligthartet al., 2021).

14.2.1 Classification of Sentiments

Classification of sentiments is one of the most frequently researched area in Computer Science. Analyzing the polarities associated with the phrases in the dataset is a sub-division of the classification of sentiments. Usually research work includes classifying the polarities as negative or positive (Wang et al., 2014). Apart from classifying the polarities as positive or negative, some of the research works include a third category called neutral. Cross-language and cross-domain classification can be considered as sub-divisions of the classification of sentiments. Ambiguity associated with the word polarities still needs to be researched. Research is carried out on models for the retrieval of information that can be considered an alternative methodology for machine learning models to implement disambiguation of word polarity (Kumar and Garg, 2020).

14.2.2 Subjectivity Classification

Subjectivity classification deals with analyzing whether subjectivity is present in the considered dataset (Kasmuri and Basiron, 2017). The prime objective of subjectivity classification is to eliminate objective phrases in the dataset. Subjectivity classification is one of the foremost tasks associated with sentiment analysis. Words that are associated with emotions like "better," "fine," etc. are detected by the subjectivity classification system (ibid.).

14.2.3 Opinion Spam Identification

Spamming of opinions is one of the major issues while implementing sentiment analysis (Krishnaveni and Radha, 2022). The advances in digitalization and growing popularity of social media networking sites, e-commerce sites and websites associated with reviews have resulted in the generation of a number of fake reviews. Fake reviews are created to either promote or demote any subject matter under consideration.

14.2.4 Detection of Language Implicitly

Sarcastic, ironic and humorous data fall under the category of implicit language. Sometimes there are situations where there is a kind of ambiguity and a sort of vagueness in certain text or speech that becomes difficult even for humans to comprehend at times. Detection of implicit language aims to tries to analyze the underlying facts associated with a certain event. Consider a phrase "I love problems", the word 'problem' is considered at factual word. The word 'love' is associated with either sarcasm or humor or it can even be ironic in nature. Conventional methods for identifying implicit language incorporate the discovery of expressions associated with laughter, usage of punctuations and emoticons.

14.2.5 Extraction of Aspects

Aspect extraction is one of the major tasks involved in sentiment analysis (Wu et al., 2022). In a particular scenario there can be a target entity, for example, the target entity can represent a product, an event, a person or an organization (Kumar and Sebastian, 2012). The concept of fine-grained sentiment analysis is quite popular (ibid.). Extraction of aspects considering blogs that do not have any specific topics and social media data like Tweets is extremely important. Several methods are used for the extraction of aspects. The conventional and the first and foremost method to implement extraction of aspects is analysis based on frequency. The analysis based on frequency involves identifying regularly incorporated compound nouns and nouns that can most probably correspond to aspects. A significant approach for frequency-based analysis for aspect extraction says that if the noun or compound noun occurs at a minimum of 1 percent of the statements, then it can be represented as an aspect. The next way to figure out the aspects is on the basis of syntax. For example, it could be analyzing the aspects that appear before a modifying adjective that is a word carrying emotions. Through this method, aspects that do not occur frequently can be discovered. An algorithm based on syntax is suggested by Qiu et al. (2009).

14.2.6 Datasets Associated with Sentiment Analysis

The contents generated by users are generally adopted by researchers to carry out sentiment analysis. The dataset considered for analysis depends on the field in which opinion mining is performed. The datasets associated with social media networking sites like Twitter and Facebook are highly subjective in nature (Valle-Cruz et al., 2022). The language adopted is informal across the social media sites, whereas, in news articles the data is highly objective and written in formal language. Vast amounts of literature are available considering Twitter as a dataset. Tweets are a concise and precise method of expressing one's opinion on any subject matter. Tweets contain rich hash tags and a number of references (Ibrahim and Salim, 2013). A number of deep learning and machine learning models can easily be implemented to carry out opinion mining of social media networking sites (Pasupa and Ayutthaya, 2022).. Tweets for carrying out sentiment analysis can easily be scraped using the snscrape library of Python. The scraped tweets contain a Tweet ID associated with each tweet, the location from where the tweets originated, the data and time associated with each tweet and several other pieces of information about the tweet. Many research articles give prime importance to Twitter data (Ibrahim and Salim, 2013). Tumblr and Facebook are also quite popular social media networking sites that are also considered suitable databases for carrying out sentiment analysis. A number of datasets pertaining to reviews related to restaurants and movies are usually considered models that are built for the classification of text. Star ratings are most probably associated with each review that facilitates the machine learning and deep learning models to be built. Polarity can be obtained using the star ratings.

14.2.7 Approaches to Sentiment Analysis

The approaches to sentiment analysis can be broadly classified into knowledge-associated methods, statistical methods (deep learning and machine learning) and hybrid methodologies that combine knowledge associated and statistical methods (Cambria et al., 2016). Opinion mining systems involve several techniques for pre-processing the dataset under consideration. Data pre-processing aims to improve the quality of the dataset by removing the inaccuracies and inconsistencies using several programming languages like R, Python, etc. A number of feature selection models like Principal Component Analysis are also incorporated.

1. *Machine learning-based approaches*: Machine learning for sentiment analysis can be carried out considering unsupervised, semi-supervised and supervised learning techniques (Amulya et al., 2022; Bhagat and Bakariya, 2022; Jorvekar and Gangwar, 2022; Shah et al., 2022).
2. *Unsupervised learning*: Unsupervised learning does not require definitions or any sort assumptions or labels to be associated with the dataset. Basically unsupervised learning technique does not need a training dataset.
3. *Semi-supervised learning*: Semi-supervised learning includes both labeled and unlabeled datasets during the process of training. The amount of effort put in by the humans can be reduced by a considerable amount by adopting a semi-supervised learning technique. Semi-supervised learning also results in a decent accuracy score. Aspect-associated opinion mining can be carried out using the semi-supervised learning mechanism. It is quite easy to obtain and capture the opinions from a sentiment-rich dataset (Nandedkar and Patil, 2022). Semi-supervised techniques are quite popular in the domain of Twitter sentiment analysis (da Silva et al., 2016).
4. *Supervised learning*: Research is carried out mainly by adopting a supervised machine learning technique that needs a labeled training dataset, without which it cannot be implemented for any tasks. Classification of tweets as positive, negative or neutral can easily be carried out using the supervised machine learning techniques. Some of the most widely used supervised models for Twitter sentiment analysis are Support Vector Machine, Logistic Regression, Random Forest, etc.
5. *Deep learning based approaches*: Machine learning has given rise to an interesting sub-field in Computer Science, i.e., deep learning. Recently, researchers have incorporated several deep learning algorithms for Twitter sentiment algorithms (Balakrishnan et al., 2022). For deep learning, the features of the dataset are obtained and learned on their own by the machine automatically, resulting in better performance and accuracy. The architecture associated with deep learning models is quite complex (Shaaban et al., 2022). Convolutional Neural Network (CNN), Deep Neural Network (DNN) and Recurrent Neural Network are some of the popular models among deep

learning (Javed and Muralidhara, 2022; Meena et al., 2022; Mohanty and Mohanty, 2022).

6. *Lexicon-based approach*: One of the most conventional approaches for sentiment analysis is by considering the lexicon-based methodology (Hemmatian and Sohrabi, 2017). The lexicon-based technique goes through the entire document analyzing the terms that involves sentiments associated with them.

14.2.8 Levels of Sentiment Analysis

1. *Aspect-based sentiment analysis*: The conventional document-based classification of sentiments aims to analyze the general emotion of a particular text and classifies the text as positive, neutral or negative. Aspect level sentiment analysis tries to figure out the emotion behind a particular aspect in a specific context.

2. *Document-based sentiment analysis*: The entire text document is considered as a whole for analysis (Wang et al., 2014). However, there are quite a few challenges in adopting document-level sentiment analysis as there could be several contradictory opinions expressed in a single document in multiple ways (Kumar and Teeja, 2012).

3. *Sentence-based sentiment analysis*: Specific statements or sentences in a particular document are considered. The polarity associated with each sentence is analyzed.

There are challenges with respect to sentiment analysis. Sentiment analysis is dependent on the domain, hence the problem of domain dependency arises. Opinion mining depends greatly on the language under consideration. Linguistic dependency is a major issue that creates problems while implementing sentiment analysis. Vast amounts of research are being carried out, using the English corpus as a dataset. Chinese and Spanish languages are also considered useful for sentiment analysis. Not much research in the field of opinion mining is carried out on other languages. Opinion mining models that try to work on datasets incorporating several languages is quite challenging (Kumar and Garg, 2019; Qazi et al., 2017). Opinion spams need to be detected (Qazi et al., 2017). Detection of opinion spams is still quite challenging in opinion mining, as fake opinions quickly spread across the web (Vosoughi et al., 2018).

14.3 Research Methodology

14.3.1 Data Extraction

Global tweets on COVID-19 were retrieved using the snscrape Python library and 500 tweets from the month of May 2020 were obtained. Keywords like #covidStrain, #covidstress, #covidanxiety, #covidmentalhealth, #coviddepression, #covidpain, #covidfear, #covidmental, #alonetogether were considered when extracting the tweets. The components of the dataset considered are tweet ID, the actual tweet, the username associated with tweets, the location from which the tweets originated, the date and time associated with each tweet. The extracted tweets consist of inconsistencies and need to undergo the step of pre-processing to eliminate the inconsistencies.

14.3.2 Data Pre-Processing

The extracted tweets are subjected to pre-processing to eliminate inconsistencies. Python programming is adopted to carry out pre-processing of the retrieved tweets. Initially, the necessary packages like numpy, pandas, re, nltk, spacy and string were imported. To maintain a sense of uniformity among the tweets, all the tweets were converted to lower case. Punctuation was removed. Some of the most commonly used words in English like "a," "but," "is" are considered as stop words. Stop words are those that are not necessary from an analysis point of view. Stop words in the considered dataset are eliminated using the nltk package. Rare words in the dataset are removed. It is quite important to obtain the root form of a word to morphologically analyze the words. Lemmatization is an important aspect of NLP that reduces a word to its root form. The nltk package of Python plays a great role in lemmatization of the considered tweets. Hyperlinks that are not necessary are also scraped out.

Emoticons and emojis that convey the emotional feelings of a person in a conversation play an important role in opinion mining. Tweets contain lots of such emoticons and emojis. The emoticons and emojis are converted to their respective meanings. A number of HTML tags present in the dataset are also eliminated since they are of no significance for analysis. A number of short forms, such as Ttyl ("Talk to you later") are used by users of social media sites like Twitter and Facebook. The short forms or chat words like Ttyl are converted to the appropriate text form. Duplicate tweets in the dataset are eliminated. Removal of duplicate tweets resulted in a total of 498 tweets.

14.3.3 Polarity Classification

The pre-processing step of opinion mining removes inaccuracies and inconsistencies in the considered dataset. The next step involves classifying each tweet as positive or negative (Kalaivani and Jayalakshmi, 2022). The process of assigning the tags to each tweet as positive or negative is termed polarity classification. The TextBlob library of Python plays an important role in carrying out several Natural Language Processing (NLP) tasks, sentiment analysis and classification procedures.

14.3.4 Tweet Classification

The present study involves distinguishing the tweets as negative or positive using a novel auto-encoder network-based ensemble technique for sentiment analysis

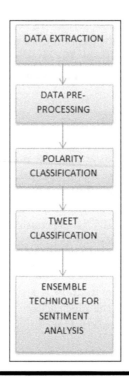

Figure 14.1   Representation of research methodology.

using tweets on COVID-19 data. Figure 14.1 gives a pictorial representation of the research methodology adopted.

14.3.5 Auto-Encoder

An auto-encoder comes under the category of unsupervised learning. Auto-encoders belong to the class of feed forward neural networks. They can also be interpreted as a self-supervised method because the technique obtains its own labels using the training dataset. The major aspects involved in auto-encoders are the construction of the input layer, the encoder network, the decoder network and the output layer. The dimensionality associated with the output layer of auto encoders is similar to that of the input layer.

14.3.6 Adam Optimization

An optimization algorithms when applied with the machine learning or deep learning algorithms yields great results. The Adam optimization algorithm, which can be considered an extended aspect of the stochastic gradient descent, finds immense applications in the field of NLP. The Adam optimizer is considered in the present research work when defining the parameters of the auto-encoder network.

14.3.7 Ensemble Techniques for Sentiment Analysis

Ensemble techniques for sentiment analysis involve consideration of several machine learning or deep learning algorithms and blend them together to result in improved accuracy rates (Mamun et al., 2022). Ensemble techniques yield more precise results than a single model under consideration (Alsayat, 2022; Kamara et al., 2022; Rabiu et al., 2022; Singh et al., 2022). The term "weak learners" which is also known as base models is used commonly in the scenario of ensemble learning techniques. Weak learners are not complicated models; they are quite simple models that act as the basis of the development of complicated models. Bagging, boosting and stacking are the three categories that fall under the umbrella of the ensemble learning technique.

- *Bagging technique*: The bagging technique, popularly represented as a bootstrap aggregation, is implemented to minimize the variation within an inconsistent dataset. In the scenario of bagging, an arbitrary portion of data involved in the training is chosen with substitution. Non-heterogeneous weak learners are taken into consideration while adopting the bagging method.
- *Boosting technique*: The boosting ensemble technique replaces the dataset considered for training, in order to give importance to illustrations that have gone wrong in the training dataset in the models that were fitted earlier. The boosting ensemble method centers on the idea of rectifying errors involved in prediction. A sequential manner is followed while fitting and integrating the models into the ensemble.
- *Stacking technique*: The stacked generalization, also called the stacking technique, is a method to ensemble several classification models. Stacking combines several classification models through the use of meta learning. In the stacking technique, a number of base models or algorithms are considered. A training dataset is considered on which the base models or algorithms are trained. The final result of all the base models is obtained and the meta model is trained on the end result of all the base models. The present research work implements the stacking technique. The base algorithms implemented are the decision tree classifier, the gradient boosting classifier, logistic regression and the genetic algorithm. The meta classifier considered is the support vector machine.

14.3.8 Decision Tree Classifier

The decision tree classifier comes under the category of supervised machine learning algorithms. They are non-parametric in nature. A decision tree algorithm finds great applications to perform regression and classification tasks. Mainly, the decision tree algorithm is used widely to perform the tasks that involve classification. The structure of the algorithm corresponds to that of a tree kind of structure. A number of nodes are associated with the tree structure. Features associated with the dataset are represented by internal nodes. Decision rules associated with the decision tree are represented by the branches of the tree structure.

14.3.9 Gradient Boosting Classifier

Gradient boosting is one of the most efficient algorithms in machine learning. Variance and bias errors are the major categories of errors in the machine learning domain. Bias error can be minimized to a great extent using the gradient boosting model. Regression and classification tasks can be performed easily considering gradient boosting. The gradient boosting algorithm plays a major role in implementing ensemble techniques.

14.3.10 Logistic Regression

Logistic regression comes under the umbrella of supervised machine learning techniques. Logistic regression is a famous model that can be implemented to carry out regression tasks. Samples of data can easily be classified using logistic regression. Logistic regression can thus be considered a classification algorithm. The end result of a categorically dependent variable can be predicted easily using the logistic regression classifier. Logistic regression is quite similar to linear regression. A logistic regression model is implemented as a part of the ensemble model. Theory of predictive modeling is implemented using logistic regression.

14.3.11 Genetic Algorithm

The term "genetic" comes from the theory of evolution in the field of biology. The idea of natural selection is brought about by the genetic algorithm. The genetic algorithm plays an important role in problems that require optimization. In the present work, GeneticSelectionCv is used. The features from the dataset can be obtained from GeneticSelectionCv. The execution of code pertaining to GeneticSelectionCv results in the display of selected features from the dataset under consideration.

14.3.12 Support Vector Machine

The support vector machine (SVM), one of the most popular algorithms for regression and classification, comes under the category of supervised machine learning techniques. A perfect boundary of decision is created by the SVM algorithm that can separate the space of n-dimension into several classes as a newly obtained bit of data can easily be placed in the right class. Boundary of decision is also termed a hyper plane. The present research work incorporates the stacking technique, as stated. SVM classifier is chosen to be the meta classifier. A well-defined training dataset is considered. The training dataset contains global tweets on COVID-19 from May 2020. The tweets in the training dataset are labeled as positive or negative. A positive tweet is represented with the digit 1 and a negative tweet is represented with the digit 0. The base-level models considered in the present research, namely decision tree, gradient boosting, logistic regression and GeneticSelectionCv, are trained, based on the well-defined training dataset formulated. The meta-classifier under consideration is SVM, which is trained on the obtained outputs of the base-level classifiers.

14.3.13 Dataset Visualization

Dataset visualization helps to obtain a systematic pictorial representation of the dataset under consideration. Interesting characteristics and patterns in the dataset can be analyzed efficiently using several graphs.

14.3.14 Word Cloud

A word cloud is one of the most fascinating dataset visualization techniques. A word cloud helps to discover interesting trends in the data. Every word in the word cloud appears in a different font size and color. Textual data is represented in a clear and succinct manner according to its frequency of occurrence. Size associated with each word in the dataset plays a major role in the analysis of word clouds. The bigger and bolder the word, the greater is the frequency of the occurrence of that particular word in the dataset.

14.4 Packages/Libraries of Python

The following are some of the libraries imported as a part of the current research work:

- *Pandas*: An open source freely available library of Python, Pandas is extremely powerful to carry out effective analysis and manipulation of data. Pandas finds great applications in the field of Data Science and analytics. The labeled dataset can quickly be processed with the flexible Python package Pandas.
- *NumPy*: NumPy is one of the most famous Python libraries to perform data analysis. NumPy supports the concept of arrays. Mathematical calculations on arrays can be efficiently performed on arrays using NumPy. Processing of matrices can easily be performed using NumPy. Effective data storage is facilitated using the NumPy library.

- *matplotlib.pyplot*: pyplot in Python comes under the category of sub-modules. The utilities of matplotlib.pyplot come under the pyplot sub module. matplotlib.pyplot plays an important role in plotting graphs that helps to visualize huge amounts of data.
- *Seaborn*: Data exploration is a major task of NLP. Interesting characteristics and patterns in the dataset are extremely important for the analysis, that can be carried out using the Seaborn library. Graphs pertaining to several statistical aspects can be plotted quite easily.
- *Nltk*: The Natural Language Toolkit (Nltk) of Python is comprised of a number of libraries. Statistical and mathematical processing of statements written in natural English can be performed using Nltk. The libraries present in Nltk are semantic reasoning, stemming, tokenization, classification, tagging and parsing.

14.4.1 mlrose: Machine Learning, Randomized Optimization and Search

Several optimization algorithms such as hill climbing, genetic algorithm and simulated annealing that find great applications in solving minimization and maximization of problems can be applied using the mlrose package of Python.

14.4.2 TfidfVectorizer

tf-idfVectorizer is a Python library. TF-IDF stands for Term Frequency Inverse Document Frequency. TF-IDF is one of the most common algorithms used by researchers across the NLP domain, TF-IDF converts the textual data into appropriate numerical representation. The conversion of textual data into its appropriate numerical representation helps in fitting any machine learning or deep learning algorithm.

14.4.3 train_test_split

train_test_split is a function in Python that comes under the sklearn model selection that helps to divide the dataset into training and testing datasets. Arbitrary partitions of the considered dataset will automatically be made by the sklearn train_test_split function.

14.4.4 Logistic Regression

Scikit-learn library of Python contains a specific pattern for modeling the machine learning algorithms. Logistic regression classifier is modeled using the Scikit-Learn library.

14.5 Conclusion

The total number of tweets considered for analysis is 500 (tweets on COVID-19 from May 2021). The number of tweets after removal of duplicate tweets is 498. Textblob library of Python was considered to determine the polarity of the tweets. The number of positive tweets in the dataset is 216 and the number of negative tweets in the dataset is 79. Comparing the numbers of positive tweets and negative tweets, the number of positive tweets is more, i.e., 216. A novel algorithm which is a culmination of machine learning, deep learning, nature inspired techniques and artificial neural networks is implemented. The suggested model is an ensemble of decision tree algorithm, gradient boosting, logistic regression and genetic algorithm based on the auto encoder technique. The objective of the research was to classify the tweets as positive or negative for the considered dataset, that is tweets on COVID-19 from the month of May 2021. The proposed novel ensemble technique resulted in an accuracy score of 81.25 percent.

Acknowledgments

Special thanks to Professor Rohini and Professor Joy Paulose for the constant support and guidance throughout this research.

References

Alali, Muath, Nurfadhlina Mohd Sharef, Masrah Azrifah Azmi Murad, Hazlina Hamdan, and Nor Azura Husin (2022) "Multitasking Learning Model Based on Hierarchical Attention Network for Arabic Sentiment Analysis Classification." *Electronics*, vol. 11, no. 8, pp. 1193–1197.

Alsayat, Ahmed. (2022) "Improving Sentiment Analysis for Social Media Applications Using an Ensemble Deep Learning Language Model." *Arabian Journal for Science and Engineering*, vol. 47, no. 2, pp. 2499–2511.

Amulya, K., S. B. Swathi, P. Kamakshi, and Y. Bhavani (2022) "Sentiment Analysis on IMDB Movie Reviews Using Machine Learning and Deep Learning Algorithms." Paper presented at 2022 4thInternational Conference on Smart Systems and Inventive Technology (ICSSIT), pp. 814– 819,

Balakrishnan, Vimala, Zhongliang Shi, Chuan Liang Law, Regine Lim, Lee Leng Teh, and Yue Fan (2022) "A Deep Learning Approach in Predicting Products' Sentiment Ratings: A Comparative Analysis." *The Journal of Supercomputing*, vol. 78, no. 5, pp. 7206–7226.

Bhagat, Meenu, and Brijesh Bakariya (2022) "Sentiment Analysis Through Machine Learning: A Review." Paper presented at 2nd International Conference on Artificial Intelligence: Advances and Applications, pp. 633–647.

Cambria, Erik, Dipankar Das, Sivaji Bandyopadhyay, and Antonio Feraco (2017) "Affective Computing and Sentiment Analysis." In Erik Cambria, Dipankar Das, Sivaji

Bandyopadhyay, and Antonio Feraco (Eds.), *A Practical Guide to Sentiment Analysis*, Singapore: Springer, pp. 1–10.

Chagas, Beatriz, Nery Rodrigues, Julio Augusto Nogueira Viana, Olaf Reinhold, Fabio Lobato, Antonio F.L. Jacob, and Rainer Alt (2018) "Current Applications of Machine Learning Techniques in CRM: A Literature Review and Practical Implications." Paper presented at 2018 IEEE/WIC/ACM International Conference on Web Intelligence (WI), pp. 452–458.

Dandekar, Aditi and Narawade, Vaibhav (2022) "Twitter Sentiment Analysis of Public Opinion on COVID-19 Vaccines." In J. C. Bansal, A. Engelbrecht, and P. K. Shukla (Eds.), *Computer Vision and Robotics*, Singapore: Springer, pp. 131–139.

Dang, C. N., Dang, M. N. Moreno-García, and F. De La Prieta (2021) "Hybrid Deep Learning Models for Sentiment Analysis," *Complexity*, vol. 2021.

Daradkeh, Mohammad Kamel (2022) "A Hybrid Data Analytics Framework with Sentiment Convergence and Multi-Feature Fusion for Stock Trend Prediction." *Electronics*, vol. 11, no. 2, pp. 250–257.

da Silva, Nádia, Félix Felipe, Luiz F.S. Coletta, Eduardo R. Hruschka, and Estevam R. Hruschka Jr. (2016) "Using Unsupervised Information to Improve Semi-Supervised Tweet Sentiment Classification." *Information Sciences*, vol. 35, pp. 348–365.

Hemmatian, Fatemeh and Mohammad Karim Sohrabi (2017) "A Survey on Classification Techniques for Opinion Mining and Sentiment Analysis." *Artificial Intelligence Review*, vol. 52, pp. 1495–1545.

Ibrahim, Mnahel Ahmed and Naomie Salim (2013) "Opinion Analysis for Twitter and Arabic Tweets: A Systematic Literature Review." *Journal of Theoretical & Applied Information Technology*, vol. 56, no. 3.

Javed, Nazura and B. L. Muralidhara (2022) "Emotions During Covid-19: LSTM Models for Emotion Detection in Tweets." In *Proceedings of the 2nd International Conference on Recent Trends in Machine Learning, IoT, Smart Cities and Applications*, pp. 133– 148.

Jorvekar, Ganesh N. and Mohit Gangwar (2022) "Multi-Entity Topic Modeling and Aspect-Based Sentiment Classification Using Machine Learning Approach." In *Proceedings of International Conference on Recent Trends in Computing*, pp. 537–547.

Kalaivani, M. S. and S. Jayalakshmi (2022) "Text-Based Sentiment Analysis with Classification Techniques: A State-of-Art Study." In S. Smys, R. Palanisamy, A. Rocha, and G. N. Beligiannis (Eds.), *Computer Networks and Inventive Communication Technologies*, Singapore: Springer, pp. 277–285.

Kamara, Amadu Fullah, Enhong Chen, and Zhen Pan (2022) "An Ensemble of a Boosted Hybrid of Deep Learning Models and Technical Analysis for Forecasting Stock Prices." *Information Sciences*, vol. 594, pp. 1–19.

Kasmuri, Emaliana, and Halizah Basiron (2017) "Subjectivity Analysis n Opinion Mining: A Systematic Literature Review." *International Journal of Advances in Soft Computing and its Applications*, vol. 9, no. 3, pp. 132–159.

Kaur, Gaganpreet, Amandeep Kaur, and Meenu Khurana (2022) "A Review of Opinion Mining Techniques." *ECS Transactions*, vol. 107, no. 1, 10125.

Kaushik, Nainika and Manjot Kaur Bhatia (2022) "Twitter Sentiment Analysis Using K-means and Hierarchical Clustering on COVID Pandemic." Paper presented at International Conference on Innovative Computing and Communications, pp. 757–769.

Krishnaveni, N. and V. Radha (2022) "A Hybrid Classifier for Detection of Online Spam Reviews." In *Artificial Intelligence and Evolutionary Computations in Engineering Systems*, pp. 329–339. DOI:10.1007/978-981-16-2674-6_25

Kumar, Akshi and Geetanjali Garg (2020) "Systematic Literature Review on Context-Based Sentiment Analysis in Social Multimedia." *Multimedia Tools and Applications*, vol. 79, no. 21, pp. 15349–15380.

Kumar, Akshi, and Mary Sebastian Teeja (2012) "Sentiment Analysis: A Perspective on Its Past, Present and Future." *International Journal of Intelligent Systems and Applications*, vol. 4, no. 10.

Ligthart, Alexander, Cagatay Catal, and Bedir Tekinerdogan (2021) "Systematic Reviews in Sentiment Analysis: A Tertiary Study." *Artificial Intelligence Review*, vol. 54, no. 7.

Mamun, Md, Mashiur Rahaman, Omar Sharif, and Mohammed Moshiul Hoque (2022) "Classification of Textual Sentiment Using Ensemble Technique." *SN Computer Science*, vol. 3, no. 1, pp. 1–13.

Meena, Gaurav, Krishna Kumar Mohbey, and Ajay Indian (2022) "Categorizing Sentiment Polarities in Social Networks Data Using Convolutional Neural Network." *SN Computer Science*, vol. 3, no. 2, pp. 1–9.

Mohanty, Mohan Debarchan, and Mihir Narayan Mohanty (2022) "Verbal Sentiment Analysis and Detection Using Recurrent Neural Network." In S. De et al. (Eds.), *Advanced Data Mining Tools and Methods for Social Computing*, New York: Academic Press, pp. 85–106.

Nandedkar, S. C., and J. B. Patil (2022) "SeAbOM: Semi-Supervised Learning for Aspect-Based Opinion Mining." In *Proceedings of International Conference on Data Science and Applications*, pp. 479–489.

Paliwal, Sushila, Suraiya Parveen, M. Afshar Alam, and Jawed Ahmed (2022) "Sentiment Analysis of COVID-19 Vaccine Rollout in India." *ICT Systems and Sustainability*, pp. 21–33. DOI:10.1007/978-981-16-5987-4_3

Pandey, A. C., Pandey, A. Kulhari, and D. S. Shukla (2022) "Enhancing Sentiment Analysis Using Roulette Wheel Selection-Based Cuckoo Search Clustering Method," *Journal of Ambient Intelligence and Humanized Computing*, vol. 13, no. 1, pp. 1–29.

Pasupa, Kitsuchart, and Thititorn Seneewong Na Ayutthaya (2022) "Hybrid Deep Learning Models for Thai Sentiment Analysis." *Cognitive Computation*, vol. 14, no. 1, pp. 167–193.

Pozzi, Federico Alberto, Elisabetta Fersini, Enza Messina, and Bing Liu (2017) "Challenges of Sentiment Analysis in Social Networks: An Overview." In F. Pozzi, et al. (Eds.), *Sentiment Analysis in Social Networks*, New York: Morgan Kaufmann, pp. 1–11.

Qazi, Atika, Ram Gopal Raj, Glenn Hardaker, and Craig Standing (2017) "A Systematic Literature Review on Opinion Types and Sentiment Analysis Techniques: Tasks and Challenges." *Internet Research*, vol. 27, pp. 608–630.

Qiu, Guang, Bing Liu, Jiajun Bu, and Chun Chen (2009) "Expanding Domain Sentiment Lexicon through Double Propagation." *IJCAI*, vol. 9, pp. 1199–1204.

Rabiu, Idris, Naomie Salim, Maged Nasser, Faisal Saeed, Waseem Alromema, Aisha Awal, Elijah Joseph, and Amit Mishra (2022) "Ensemble Method for Online Sentiment Classification Using Drift Detection-Based Adaptive Window Method." In *International Conference on Reliable Information and Communication Technology*, pp. 117–128.

Ruz, Gonzalo A., Pablo A. Henríquez, and Aldo Mascareño (2022) "Bayesian Constitutionalization: Twitter Sentiment Analysis of the Chilean Constitutional Process through Bayesian Network Classifiers." *Mathematics*, vol. 10, no. 2, pp. 166–172.

Shaaban, Mai A., Yasser F. Hassan, and Shawkat K. Guirguis (2022) "Deep Convolutional Forest: A Dynamic Deep Ensemble Approach for Spam Detection in Text." *Complex & Intelligent Systems*, vol. 8, no. 6, pp. 4897–4909.

Shah, Parita, Priya Swaminarayan, and Maitri Patel (2022) "Sentiment Analysis on Film Reviews in Gujarati Language Using Machine Learning." *International Journal of Electrical and Computer Engineering*, vol. 12, no. 1.

Shukla, Priyanka, and Adarsh Garg (2022) "Sentiment Analysis of Online Learners in Higher Education: A Learning Perspective through Unstructured Data." In S. Pathak et al. (Eds.), *Intelligent System Algorithms and Applications in Science and Technology*. New York: Apple Academic Press, pp. 157–170.

Singh, Vishakha, Sameer Shrivastava, Sanjay Kumar Singh, Abhinav Kumar, and Sonal Saxena (2022) "StaBle-ABPpred: A Stacked Ensemble Predictor Based on Bilstm and Attention Mechanism for Accelerated Discovery of Antibacterial Peptides." *Briefings in Bioinformatics*, vol. 23, no. 1.

Valle-Cruz, David, Vanessa Fernandez-Cortez, Asdrúbal López-Chau, and Rodrigo Sandoval-Almazán (2022) "Does Twitter Affect Stock Market Decisions? Financial Sentiment Analysis During Pandemics: A Comparative Study of the HLNL and the Covid-19 Periods." *Cognitive Computation*, vol. 14, no. 1, pp. 372–387.

Vosoughi, Soroush, Deb Roy, and Sinan Aral (2018) "The Spread of True and False News Online." *Science*, vol. 359, no. 6380, pp. 1146–1151.

Wang, Gang, Jianshan Sun, Jian Ma, Kaiquan Xu, and Jibao Gu (2014) "Sentiment Classification: The Contribution of Ensemble Learning." *Decision Support Systems*, vol. 57, pp. 77–93.

Wu, Haiyan, Zhiqiang Zhang, Shaoyun Shi, Qingfeng Wu, and Haiyu Song (2022) "Phrase Dependency Relational Graph Attention Network for Aspect-Based Sentiment Analysis." *Knowledge-Based Systems*, vol. 236, 107736.

Wu, Jian, Xiaoao Ma, Francisco Chiclana, Yujia Liu, and Yang Wu. (2022) "A Consensus Group Decision Making Method for Hotel Selection with Online Reviews by Sentiment Analysis." *Applied Intelligence*, vol. 52, no. 9, pp.1–25.

Chapter 15

Economic Sustainability, Mindfulness, and Diversity in the Age of Artificial Intelligence and Machine Learning

Ranjit Singha and Surjit Singha

15.1 Use of AI and ML in Traditional Industry

In India, under the Company Act, 2 per cent of corporate social responsibility (CSR) funding is mandatory, so it is a boost for many non-governmental organizations (NGOs) to receive such CSR funding from various organizations to develop their operations in society. Often, social work is motivated by emotion, and in that spirit, many NGOs step forward to save traditional industries and ancient technology. And then they fail to recognize that time has passed and new technology has been introduced. Today, most NGOs in India are working to save traditional industries. They believe that they will keep the tradition alive; in the same spirit, those involved in such manufacturing traditions want to keep it alive. Understandably, people have strong attachments to traditional skills, technology, culture, practices, and machines that have been in use for a long time. Stories have been created over the ages; it's not easy to leave the past. At the same time, the new generation is moving to cities because they have seen that the old traditional business is complex and does not pay well. In

DOI: 10.1201/9781003424550-15

all this, the capitalist can use automation and continue the traditional business in an automated manner and make a profit. One such industry is the handloom-based industry, pottery making. However, it is widely assumed that automation cannot truly replace manual labour. If we consider the fabric's durability, this is most likely correct. A hand-made cloth is much more durable and can last a lifetime, but today's consumers are looking for good looks, low prices, and constant fashion changes. Creating a hand-made fabric takes a lot of labour, enthusiasm for the product, love for the customer, and attention to detail. Automated technology has replaced traditional weaving, and similar fabrics can be made in an hour instead of a week. This is a problem for weavers because machine-made cloth is less expensive and can be more beautiful than hand-made cloth; the only difference is the durability of the cloth; machine-made cloth will not last as long as hand-made cloth. And customers may not understand the distinction between machine-made and hand-made cloth. With the advent of technology, there are ways that the automated cloth manufacturer will be able to sell their products, even if there are some legal restrictions on selling automatic machine-made cloth in some geographical areas.

The vicious cycle that the weavers suffer does not tend to stop because there is too much exploitation in the whole process. Today's social climate has deteriorated where a weaver who maintains traditional looms faces difficulty finding a suitable bride or groom because the sustainability of that career is in doubt. Such social issues can be resolved by incorporating technology into the entire system; production will increase if traditional looms can be automated. Thus, a lower-priced product can be sold; however, money will be required to implement such technology. Many looms are automated; the owners provide houses and looms; skilled professionals work as labourers and are paid per sari basis. The problem with modern technology is that the poor can learn the technology and have the ability to gain the skills, but they cannot afford to purchase the technology. Most weavers in India are unaware of intellectual property rights (IPR) and do not have their designs' copyright or design patents. Most of their designs are copyright filed, the design patent holders are the owners of large corporations, leaving weavers with no choice but to work with the same design that they are familiar with, for someone else. Even if weavers are given the technology and skills, they do not have the legal right to create the same design. The solution is for them to know how to file copyright, patents, and other legal aspects of the entire process or for someone to assist them in the whole process. The filing of copyright, trademarks and patents in India is a complicated process that needs simplification for the weavers. Obtaining a digital signature could be automated using the PAN card number or AADHAAR card number, thereby reducing the cost to purchase a separate digital signature. This legal process of obtaining a digital signature, patent drafting, and copyright filing may require legal professionals, but all of this can be done individually if NGOs, government bodies, or institutions provide real-time training in patent drafting and copyright filing.

In most cases, NGOs and government agencies wash their hands of the matter simply by raising awareness and submitting a report. This doesn't help, and most of the success stories submitted by various agencies are fabricated. The process of copyright filing and patent filing can be automated in a simplified manner so that any citizen should be able to file a copyright or patent.

15.2 Use of AI to Create an Agricultural Database

One of the pressing challenges in India pertains to the absence of a comprehensive national database encompassing essential information on land use, soil health, crop patterns, disease prevalence, and agricultural extension services. The landowners are having difficulty finding agricultural labour and water for the soil. Some landowners converted their properties into multi-story buildings, schools, and colleges. In most cases, the land's viability for cultivation was finished, and farmers are familiar with one type of cultivation. That is main issue with tobacco cultivation. At present, tobacco cultivation in India is declining not as a result of any livelihood training but rather as a result of decreased land vitality for the tobacco product. A farmer who knows rice cultivation is generally familiar with the entire process. However, they are concerned about shifting cultivation to other crops because they are unfamiliar with the whole process and the markets. For Indian agriculture to be sustainable, there is an urgent need for scientific intervention. There are agricultural offices in every district in India, and the government employs qualified professionals in each of these offices. However, much of the work is not being done as it should be. The potential solution is an interactive app that understands the local language and answers all of the farmer's questions in the local language. The challenge is obtaining the information from the local language database, which must exist. It would take some time for such an interactive app to learn most aspects of agriculture. Ideally, the learning algorithm of such technology can be automated, which has its own risk of providing incorrect information. The best answer is a human teacher, but that takes time and money, and it is the best way to go. In the meantime, such a local language database on agriculture must be built, which is a costly endeavour, given that India has 1,652 or more languages. Koch languages are on the verge of extinction; there are opportunities to preserve many languages using machine learning and AI and teach the same language with the same intensity to non-native populations from other parts of the world. We can preserve a great civilization and avert its extinction by using technology.

15.3 Use of AI to Be Sensitive to Users' Emotions

With the aid of virtual reality, technology is now capable of eliciting users' real-time emotions and feelings. By leveraging the entire process of visualization, reading, and listening with AI, machine learning, and beyond, the digital future has the potential to incorporate real-time emotions and feelings. This would entail emotional responses on both ends and a variety of other technologies and users, but each of these carries its own set of risks. Through AI and ML, it is possible to cultivate mindfulness and an awareness of diversity and communal harmony, as AI and ML are capable of inferring the emotional and cognitive states of the people with whom they interact. Artificial intelligence and machine learning could be used to educate weavers and farmers about their legal rights, cultivation methods, banking processes, and the harmful effects of tobacco use and other health-related issues.

15.4 Livelihood Training and a Caution for CSR Funders

With the sponsored livelihood training offered by corporate social responsibility (CSR) funding, the growing concern is with unsolicited training by unprofessional and unskilled trainers at the local level; CSR funding is being misused unknowingly due to a lack of priority for quality specific to training. In the first place, it appears that livelihood training is the answer to eradicating poverty. However, it may make capitalists increase their labour power, as this is where most educational institutions in India have been heading. With the advent of AI, there would be less requirement for skilled labour, the number of such skilled labourers in a process will eventually be reduced, and specific aspects of the work will be taken care of by the AI. This is a boost for the capitalist. Promoting skilled labor for capitalists seems a poverty solution, but it leads to an increased labor pool for industry needs. Industries should recruit and train workers, avoiding CSR-funded outsourcing, and use internal funds for training, fostering self-sustaining skill development. Keep in mind that the industry is now armed with AI and ML. Some say it is obsolete and redundant to spend money and time on such training. It's more like spending money for the sake of spending, because it's required by the Company Act to devote cash to CSR-related work. It's just another activity, just for the sake of the Company Act. One of the core issues is that the country is not progressing; the poor are becoming poorer, and the rich are becoming more prosperous. Considering the tobacco industry, the farmers, the tobacco growers, the biri makers, and the cigarette rollers are poor, they suffer from health issues, and lack the financial resources to receive treatment. They are impoverished. Even though tobacco is expensive in India, tobacco leaf is referred to as "golden leaf". Tobacco profits are concentrated in the hands of capitalists; only six companies have a monopoly on tobacco manufacturing cigarettes in India.

While doing CSR funding, organizations must exercise caution. Specific trades, such as making agarbatti, also known as incense sticks, in India, biri making, and cigarette rolling should never be funded, because they are carcinogenic by nature, pollute the environment, and are harmful to people's health. However, it has been observed that many top-level funding agencies in India are funding training in making agarbatti, and there are a good number of NGOs associated with such activity. Agarbati making may appear to be a solution for a short time, but it is harmful in the long run; so far, in our observation, no funding agency is found to be sponsoring Biri making or rolling cigarette training.

Some trades will be rendered obsolete by automation, and incorporating skill training for such businesses into livelihood training is a waste of time because the product will not maintain a market share due to price competition. Automation combined with AI and ML will result in a higher quality product with a better design in less time and at a lower cost. As a result, labour costs will be reduced, as will the product price. It may appear that funding such traditional trades will benefit conventional skilled workers, but this is not the case until and unless they incorporate automation, AI, and ML into their entire process. During the COVID-19 pandemic, people's purchasing power decreased; they are looking for high-quality products and better designs at lower prices.

Before providing any CSR funding, the sponsor should screen to see if an NGO or institute is involved in agarbatti making livelihood training or if the NGOs are providing skill training in specific trades that are already being automated. In such a case, no organization should provide CSR funding. The hidden reality of the time is that industries are looking for free training through NGOs to get a skilled labour force. The industry is attempting to solve its problem of skilled labour force scarcity by receiving professional labour-free training through CSR funding because this saves them money on training. While resolving the industrial problem of a lack of skilled labour will undoubtedly provide a nominal wage-based income for an individual or family, this is not the ultimate solution. A process-based training may not help because, often, the process and product are patented, design is copyright or a registered design. So, despite being highly skilled for an extended period in a process, that individual may be unable to leave that employment and start his/her own venture in the same trade in which they had been employed. Integrating skill-based training about new technology into training is a solution; however, adopting such technology for real-time manufacturing for cottage-based industry is impossible until and unless such technology is provided with a high subsidy.

15.5 Organic Farming and Sustainable Livelihoods

Organic farming can be the next sustainable livelihood because it preserves the vitality of the soil for longer. Organic farming is used in traditional farming in India, but productivity is lower, and yields are not up to the quality of those who use chemicals. As a result, there is a gradual shift in Indian agriculture away from organic to chemical-based cultivation. Farmers are shifting from indigenous varieties of seeds to hybrid-based seeds. Today, farmers use drones to spread pesticides; however, this is not feasible for all farmers; it requires investment, and farming with a tractor is difficult for many. However, drones and tractors are now available for rent, and thus many farmers are using modern technology. There is a lack of understanding of contemporary technology and scientific methods in organic farming. The region of Himachal Pradesh is well known for its organic farming because chemical-based farming is prohibited by law in that state. J&K is famous for its apple production; however, it is less well known that apple production is also possible in Arunachal Pradesh. Commercial cultivation of apples is not common in Arunachal Pradesh; however, the geographical area has the potential for commercial apple cultivation. AI can harvest, identify different pests, plant diseases, and poor farm nutrition, detect weeds, and spread herbicides around them. In India, the purchase of agricultural technology is subsidized; however, farmers are unaware that such technology exists.

Implementing modern technology is doable and affordable due to available subsidies. One of the main issues is a lack of electricity in India's remote areas. Most high-end technology requires electricity to charge the battery. If such technology were compatible with solar power, it would be possible for Indian villages to apply such technology. There is a possibility of raising awareness for such a programme. Since the early 1990s, the Government of India has used television to broadcast various farmer-related programmes, including agricultural news. There is still room to raise awareness about modern technology in agriculture. India is a country whose entire climatic condition is influenced by the Himalayan region and deserts; thus, different climatic conditions exist across India's diverse geography. The problem is how to deploy AI with 100 per cent accurate output because each region would require a different database according to the climatic conditions there. There would be variations in the pattern of plant diseases. Such a database, specifically on a regional basis, does not exist in India; we have to depend upon an international database for prediction, which might not be 100 per cent accurate, considering the varieties of crops and climatic conditions in India. This raises the question of investment in such technology, and if the output has lots of error, it would be a waste of time, money and bring the risk of wrong diagnosis and treatment. There is a need for an Indian regionally-based database of different plant species, images of diseases, pathogens, and pests. The regional-based database must be built and expanded over time. It requires a dedicated R&D team, technology, investment, and significant funds.

15.5.1 Agarbatti Smoke and Tobacco Smoke

Agarbatti is also known as an incense stick. Both agarbatti smoke and tobacco smoke are harmful to one's health. The manufacturing processes of making cigarettes or bidi and incense sticks is detrimental to one's health. Agarbatti is carcinogenic by nature and is equally harmful to one's health as cigarette smoke. Even if agarbatti is made from natural ingredients, the combustion process emits carbon into the atmosphere; therefore, agarbatti in any form should be avoided (Višić et al., 2018). Incense sticks are associated with occupational hazards and indoor air pollution (Lin et al., 2008). CO, CO_2, NO_2, SO_2, and other gases are produced when incense is burned.

In addition to volatile organic compounds such as benzene, toluene, and xylenes, incense burning produces aldehydes and polycyclic aromatic hydrocarbons (PAHs). Air pollution is significant in all areas where agarbatti burning occurs, hazardous to one's health. Making incense sticks out of waste flowers may appear to be a solution in many cases, but it is not; it has the same potential to pollute the air (Gallego et al., 2011). The best way to manage flower waste is to dispose of it in the soil and convert it into plant manure. Making agarbatti from waste flowers is the wrong solution; instead, it adds to the existing problem. The fragrance of the agarbatti releases volatile organic compounds (VOCs), which are hazardous to health (DiGiacomo et al., 2018). Environmental tobacco smoke is harmful to health. It is associated with cardiovascular disease. The only possible difference between agarbatti and tobacco smoke is that tobacco smoke causes addiction because of nicotine, whereas agarbatti smoke may not cause any addiction.

There are smart cities across the globe that were built using Deep Learning and AI to detect pollution. CNN-LSTM (CNN Long Short-Term Memory Network) is already deployed in China to detect such pollution; there are possibilities to deploy a hybrid CNN-LSTM multivariate, which will eventually predict and depict more accurate output. Deep Learning (DL) algorithms may include Long Short-Term Memory (LSTM), Bidirectional LSTM (Bi-LSTM), Gated Recurrent Unit (GRU), Bidirectional GRU (Bi-GRU), Convolutional Neural Network (CNN), and a hybrid CNN-LSTM model, Denoising Auto-Encoders (DAE). In an ideal scenario, the LSTM model is observed to be more accurate than that of the DAE model because of the existence of more layers of LSTM. LSTM is a type of recurrent neural network (RNN) that addresses the vanishing gradient problem, enabling the network to retain long-term dependencies in sequential data. Bi-LSTM extends the LSTM model by processing the input sequence in both forward and backward directions, allowing the network to capture information from past and future contexts. GRU is another type of RNN that is similar to LSTM but has a simpler architecture. It also addresses the vanishing gradient problem and can capture long-term dependencies. Similar to Bi-LSTM, Bi-GRU processes the input sequence in both directions and captures information from both past and future contexts. CNN is a deep learning model commonly used for image processing tasks. It uses convolutional layers to automatically learn hierarchical representations from input data. This model combines the strengths of CNNs and LSTMs by using CNN layers for feature extraction from input data and LSTM layers for sequential processing and capturing dependencies, known as a Hybrid CNN-LSTM model. DAE is a type of unsupervised learning model that learns to remove noise from input data. It consists of an encoder network that maps input data to a latent space representation and a decoder network that reconstructs the original data from the latent representation. These DL algorithms encompass a wide range of models and there are many other variations and architectures available, depending on the specific problem and data characteristics.

However, it's hard to train, as it consists of more layers; any addition of layers ideally improves the accuracy of the results. The same methods can be used in India to predict better output. India is moving towards smart cities and villages, but for the time being, it is limited to road connectivity, having electricity in every home, and solar street lights. India had launched Digital India and Make in India based on its own earlier concept of Make in India initiatives. In Make in India, the idea is to collaborate with technology from various parts of the world and assemble products in India. However, the issue of technology compatibility affects the whole process, which has its own merits and demerits. AI and ML could be used to raise awareness about the detrimental effects of tobacco use and other health-related issues associated with the use of agarbatti and other pollution control mechanisms.

15.6 The Role of Banks and Financial Institutions

The bank plays a vital role in inspecting and verifying if an organization is established through CSR funding. A secured loan indicates a strong foundation. Successful training is validated by the sponsoring organization of the CSR fund, with documentation showing trainees obtaining loans. However, beyond validation, many trainees are motivated to take loans due to financial pressure related to their daughters' marriages and awareness of available subsidies, this eventually helps them, and the rest return, and many times, they fail as well, and the bank suffers from an increase in NPAs (non-performing assets). NPAs are also known as non-performing loans (NPL) or bad loans, and refers to loans or advances provided by banks and financial institutions that have stopped generating interest income or principal repayment. In other words, an NPA is a loan or credit facility on which the borrower has failed to make timely payments of interest and principal for a specified period, typically 90 days or more. When a borrower defaults on loan payments, the loan is classified as a non-performing asset by the lending institution. NPAs can arise due to various reasons, including financial distress of the borrower, economic downturns, inadequate cash flows, or mismanagement. Banks and financial institutions categorize their loans into different classifications based on the degree of repayment delays and the likelihood of recovery. The classification systems may vary across countries, but generally, loans are categorized as Substandard, Doubtful, and Loss Assets, depending on the severity of the default and the prospects of recovery. NPAs pose significant challenges to banks and financial institutions as they impact profitability, asset quality, and overall financial stability. Lenders may need to make provisions for these bad loans, setting aside funds to cover potential losses. Resolving and reducing NPAs require effective loan recovery mechanisms, restructuring options, and appropriate risk management practices to minimize their impact on the financial institution's balance sheet. Many banks provide CSR funding with the expectation that the trainee will take out a loan, which is a high-risk criterion for the bank because the trainee is new to the venture.

Furthermore, banks must exercise caution when funding cigarette- or agarbatti-related businesses, as these ventures are not sustainable for the environment and are harmful to health. Banks in India have been automated since Independence. The nationalization of India's banks can be traced back to the preamble of the Indian Constitution, which speaks about a democratic republic. Banking in India has become much smarter; its payment gateway, RuPay, the Bharepe QR code, has integrated most wallet payments into one. This was a good solution, especially during the COVID-19 crisis, lockdown, and the demonization of India. However, demonization did not deter people from using the queue because it required a real-time exchange of old currency for new. The digitization of the banking processes facilitated smooth banking during the demonetization period. Still, demonetization impacted the Indian economy due to a lack of proper systematic management of the entire process. AI can now maintain banking records much more efficiently; banking analytics are handled by AI, providing banking operations with a comprehensive understanding of the whole process. In the entire banking process, AI enables additional features such as digital payments, AI bots, and biometric fraud detection systems. Artificial intelligence and machine learning could be used to educate weavers and farmers about banking processes and to inform the general public about the banking system and processes.

15.7 Diversity

India is a diverse country with over 1,652 languages and dialects, and there may be more languages in India that are not yet known. Any CSR funding should not undermine the country's existing diversity, and all training activities must consider diversity as an essential component of training. There should be diversity among the trainees, including males, females, and another gender. It must consider all ethnicities, religions, and cultures. This could potentially become one of the elements of success in any venture because united we are strong, and divided we fall. However, various NGOs focus on empowering only one specific community, religion, or gender because the goal of establishing such is focused on the growth and development of a particular district. And it is possible that they might make a mistake by ignoring diversity and its repercussions. If they avoid diversity in training, there is a risk of lacking creativity and innovation throughout the process. Deploying AI can resolve racial discrimination; in the process, the first level of initial blockage can be removed by deploying AI. There is a possibility of discrimination based on race, colour, gender, nationality, religion and other diversity factors, but there are no such conditions with AI, provided the technology is programmed appropriately. There are opportunities for rapid and impartial assessment of organizational diversity using deep neural networks by deploying AI. It is possible to cultivate mindfulness, and an awareness of diversity and communal harmony through the use of AI and machine learning as AL and machine learning are capable of inferring the emotional and cognitive states of the people with whom they interact. By leveraging the entire process of visualization, reading, and listening with AI, machine learning, and beyond, the digital future has the potential to incorporate real-time emotions and feelings. This would entail emotional responses on both ends and a variety of other technologies and users, but each of these carries its own set of risks (Suzuki, 2019). Like cognitive robots or avatars, embodied AI is more suited to tactful conduct than disembodied AI such as chatbots or pattern categorization systems. They lack the touch surface (interestingly, "contact" has the same etymology as tact) for rich and caring interactions. Because this interactional contact surface is bidirectional, tact has both a perceptual and active component. Tact is not present when someone is aware that their actions may offend others but chooses to carry them out. So, tact must have an active component in addition to its perceptual one. In other words, tact is a practical intelligence that directs conduct to be politer by being aware of others and the current context. However, the Japanese cultural perspective permits it to be understood as a unified idea, comparable to their word *mi* which concurrently connotes the physical body, one's state of consciousness and dynamic social ties (Suzuki, 2010). Tact is an impact, effect, or sensation (Huang and Nishida, 2013). Tact is the primary measure of harmony. An essential goal of harmonizing AI is to make intelligent systems respectful of people. Virtual and physical smart assistants must know when to communicate with humans and when to disengage and keep silent. One example is an attentive quiz agent who "observes" human group interaction and avoids speaking when members are actively engaged in debate (Suzuki, 2019). Perception extends beyond the standard AI sense-think-act loop. A tactful individual can combine "all five senses into a united totality" to gain a thorough grasp of the current situation and the persons involved. Since it is agent-specific and geared towards the most appropriate physical response to a given crisis, this process of perceptual integration might be conceived of as corporeal intelligence.

Harmony and its relationship to artificial intelligence are a worthwhile goal to pursue. The Chinese Academy of Science supported Yi Zeng's creation of the Harmonious Artificial Intelligence Principles. Huawei's new operating system, which has the potential to become a national operating system, has been given the name "HarmonyOS".

Yoh'ichi Tohkura envisioned a "Convivial Society of Harmony between Humans and AI", which was enshrined in the Japanese government's 5th Science and Technology Basic Plan on "Society 5.0".

We must be aware of artificial intelligence's mediating role in our sociotechnical society. We should work hard to create artificial intelligence systems that will allow us to live in greater harmony with one another. AI harmonizes us by mediating our interactions.

15.8 Mindfulness

Being mindful could be the possible solution to most of the issues; in the first place, burning agarbatti is harmful to one's health, and many people are unaware of it. Agarbatti has never been labelled as dangerous to health. Millions of people indeed depend upon the agarbatti industry, and it's profitable; it's also associated with faith and religion (Tang et al., 2013). Recovery from tobacco addiction is possible by applying brief mindfulness (Ehrlich, 2015), and the practice of mindfulness can create mindful leaders, whereby their decisions could be conscious decisions, which could eventually help make a better policy (Wamsler et al., 2018). Mindfulness is considered a sustainable science (Wilson et al., 2019). In mindfulness, practising self-compassion could be much more helpful in the whole process, so integrating self-compassion into mindfulness could be a practical solution. AI and ML can play a role in the guided meditation process. ML can learn the likings and dislikes of the tone of sound, voice, preference of the type of voice, and sound. AI can deliver the guided meditation in the appropriate style required by the user, making any intervention more effective. There is a potential danger in this process because once AI learns the preferences of voice and sound, it can deliver any content it wants. There is a possibility of influencing any person in the world by making things happen accordingly. Artificial intelligence and machine learning could be used to educate about mindfulness and conduct mindfulness-based sessions. Web, Android, and iOS apps now provide guided meditation, and mindfulness; additional biofeedback can be tracked through artificial intelligence and machine learning. AI and machine learning can create harmony in music, arithmetic, and art. For instance, the musical notes such as C and G sound harmonious when combined because their ratio is a simple fraction, $G/C = 392 \text{ Hz}/262 \text{ Hz} = 3/2$ (Berberich et al., 2020). Verbeek (2005; 2011; 2016) developed the concept of mediation theory in technology. Additionally, a blue-green deployment model can be used to maintain two independent infrastructures or duplicate feature stores. The blue-green deployment model is a software release management strategy that aims to minimize downtime and risk during the deployment of new features or updates to a production environment. In this model, two independent infrastructures, referred to as "blue" and "green", are set up. It may also be defined as a single production environment for feature and prediction requests.

15.9 Conclusion

Human diversity and mindfulness are necessary for the sustainable implementation of artificial intelligence (AI) and machine learning (ML). This chapter has explored the various ways in which AI and ML can collaborate with humans to advance society. These include assisting with copyright and design patent registrations, cultivating mindfulness, educating weavers and farmers on their legal rights and agricultural practices, facilitating banking procedures, and promoting health-related awareness.

Moreover, by leveraging their analytical capabilities, AI and ML have the potential to enhance creativity in disciplines such as music, mathematics, and the arts. In conjunction with the cultivation of mindfulness and diversity, the blue-green deployment model can play a significant role in maintaining independent infrastructures and feature repositories.

By integrating real-time emotions and sentiments through the use of AI, ML, and other emergent technologies, the digital future holds vast potential. This would necessitate not only capturing visceral responses, but also incorporating a variety of user perspectives and technologies. The ability of AI and ML to infer emotional and cognitive states can contribute to the promotion of mindfulness, awareness of diversity, and communal harmony.

To ensure a sustainable future, it is essential to emphasize the ethical and responsible use of AI and ML, along with human values and mindfulness. By leveraging the potential of these technologies, we can improve various facets of society while fostering well-being, creativity, and a greater comprehension of human emotions and experiences.

The potential for AI and ML to contribute to societal development in India is extensive, but they must be approached with caution and honesty. CSR funders should exercise caution to ensure that funds for livelihood support are used to enhance skills and integrate automation responsibly, as opposed to replacing traditional practices. Trainers in skill development should have the necessary knowledge, credentials, and experience. Despite the fact that AI and ML have already revolutionized numerous fields, it is essential to prevent their unethical use. By establishing a balance between technological progress and ethical considerations, technology can become a potent instrument that serves the requirements of society while preserving cultural values.

Legislation should be instituted to prevent unscrupulous practices and preserve technological equilibrium. The digital future provides opportunities for the incorporation of real-time emotions, the promotion of mindfulness, diversity awareness, and social harmony. AI and ML can aid in the comprehension of emotional and cognitive states, allowing for the facilitation of guided meditation, mindfulness practices, and biofeedback monitoring for enhanced well-being. In addition, the blue-green deployment model provides a practical method for managing separate infrastructures and feature repositories.

A sustainable and beneficial future depends on the ethical and responsible application of AI and ML in accordance with human values and awareness. This strategy ensures that technology serves as a catalyst for societal advances, while respecting traditional values and promoting overall well-being.

References

Berberich, N., Nishida, T., and Suzuki, S. (2020) "Harmonizing Artificial Intelligence for Social Good." *Philosophy & Technology*, 33(4): 613–638. https://doi.org/10.1007/s13347-020-00421-8

DiGiacomo, S. I., Jazayeri, M., Barua, R. S., and Ambrose, J. A. (2018) "Environmental Tobacco Smoke and Cardiovascular Disease." *International Journal of Environmental Research and Public Health*, 16(1): 96. https://doi.org/10.3390/ijerph16010096

Ehrlich, J. (2015) "Creating Mindful Leaders and Organizations." *People & Strategy*, 38(3). https://link.gale.com/apps/doc/A425111993/AONE

Gallego Piñol, E., Roca Mussons, F. J., Perales Lorente, J. F., and Guardino Solà, X. (2011) "Simultaneous Evaluation of Odor Episodes and Air Quality in Urban Areas by Multi-Sorbent Sampling and TD-GC/MS Analysis." *Reporter*, 48: 3–5. http://hdl.handle.net/2117/15563

Huang, H., and Nishida, T. (2013) "Evaluating a Virtual Agent Who Responses Attentively to Multiple Players in a Quiz Game." *Journal of Information Processing*, 8(1): 81–96. https://doi.org/10.11185/imt.8.81

Lin, T., Krishnaswamy, G., and Chi, D. S. (2008) "Incense Smoke: Clinical, Structural and Molecular Effects on Airway Disease." *Clinical and Molecular Allergy*, 6. https://doi.org/10.1186/1476-7961-6-3

Suzuki, S. (2010) *Takt in Modern Education*. Münster: Waxmann.

Suzuki, S. (2019) "Etoku and Rhythms of Nature". In J. R. Resina and C. Wulf (eds) *Repetition, Recurrence, Returns*. Lanham, MD: Lexington Books, pp. 131–146.

Tang, Y., Tang, R., and Posner, M. I. (2013) "Brief Meditation Training Induces Smoking Reduction." *Proceedings of the National Academy of Sciences of the United States of America*, 110(34): 13971–13975. https://doi.org/10.1073/pnas.1311887110

Verbeek, P. P. (2005) *What Things Do: Philosophical Reflections on Technology, Agency, and Design*. Philadelphia, PA: The Pennsylvania State University Press.

Verbeek, P. P. (2011) *Moralizing Technology: Understanding and Designing the Morality of Things*. Chicago: University of Chicago Press.

Verbeek, P. P. (2016) "Toward a Theory of Technological Mediation: A Program For Post Phenomenological Research." In J. K. Berg, O. Friis and Robert C. Crease (eds), *Technoscience and Post Phenomenology: The Manhattan Papers*. London: Lexington Books, pp. 189–204.

Višić, B., Kranjc, E., Pirker, L., Bačnik, U., Tavčar, G., Škapin, S. D., and Remškar, M. (2018) "Incense Powder and Particle Emission Characteristics During and After Burning Incense in an Unventilated Room Setting." *Air Quality, Atmosphere & Health*, 11(6): 649–663. https://doi.org/10.1007/s11869-018-0572-6

Wamsler, C., Brossmann, J., Hendersson, H., Kristjansdottir, R., McDonald, C. P., and Scarampi, P. (2018) "Mindfulness in Sustainability Science, Practice, and Teaching." *Sustainability Science*, 13(1): 143–162. https://doi.org/10.1007/s11625-017-0428-2

Wilson, A. F., Mackintosh, K., Power, K., and Chan, S. W. Y. (2019) "Effectiveness of Self-Compassion Related Therapies: A Systematic Review and Meta-Analysis." *Mindfulness*, 10(6): 979–995. https://doi.org/10.1007/s12671-018-1037-6

Chapter 16

Adopting Streaming Analytics for Healthcare and Retail Domains

G. Nagarajan, Kiran Singh, T. Poongodi, and Suman Avdhesh Yadav

16.1 Introduction

Streaming analytics is the implementation of analytics in streaming real-time data for crucial applications that need real-time information for decision-making. In most streaming analytic applications, the transmission has minimal latency; let it be less than 1 millisecond, but it is never a zero-latency transmission [1]. The strength of the system is revealed with the precision of the real-time information that is collected. Streaming analytics is based on the theory of *streaming data processing* that processes that data in real time, one unit at a time; and *low-latency analytics* that delineate the internal events as they occur. Such processes must be continuous, immune to technical glitches and capable of reconstructing the data transfer if interrupted by failure of either internal nodes or processes.

Healthcare is an important example of applications of streaming analytics. The healthcare databases process massive heterogeneous data in real time that is being shared on an extended network. This database is used by hospitals, medical practitioners, researchers, insurance companies, government authorities and many others. Healthcare industries have grown exponentially after implementing Big Data analytics [2]. Researchers' emphasis on inspection of data stipulates identifying diseases on the basis of modifying physiology. Understanding the relations, interconnection and variations of the modality of healthcare data is crucial for streamline analytics. The latest technologies are contributing to capture the vast amounts of diversified data in real time. Researchers are focusing on this complex diversified real-time data with prime concern on their taxonomy.

The application of streamline analytics in healthcare can be exhibited in three areas: (1) image processing; (2) signal processing; and (3) genomics. These three areas show how popular among researchers streaming analytics are, where the real-time data is being processed in healthcare industries. In the healthcare industries, medical diagnosis relies on medical images such as CT, MRI, X-rays, ultrasounds, mammography. These images are the primary source of diagnosis of critical diseases, their treatments and cures. These images range from a few MB to thousands of MBs for each observation and also require huge durable storage capacity. High speed faultless algorithms and a stable storage facility are two major challenges when processing these real-time images.

Similar to medical images, physiological signals also demand fast algorithms and an efficient database. These signals are very complex in nature and are collected from the monitoring devices and sensors connected to the patient's body. Healthcare systems employ several body sensors and other monitoring devices connected through IoT systems that generate alerts and transfer patient data in real time to data servers. There is a need to develop approaches that best acquire and process the clinical data and transmit them to the servers in real time with minimal losses.

Genome sequencing is working on identifying new drug designs by analysing databases and sequencing tools to acquire all the clinical data, applying data mining algorithms and modelling the data for crucial decision-making [3]. The data required for genome sequencing is of high density and needs to be transmitted in real time. Thus, Big Data analytics are required that can process such complicated medical data. The expeditious applications of real-time data processing in healthcare result in more and more use of IoT and Big Data analytics.

Streaming analysis has made it possible for enterprises to inspect and analyse the data in real time when it is being streamed to its applications. In retail industries, the real-time data is streamed, starting right from the POS systems installed at the stores to the transactions between client and merchant bank accounts. The warehouses are also updated in real time to set the inventory levels and, following that, the supply chain management system is run. This real-time streaming of retail data helps to identify the purchasing trends and recommend better product specifications. Such analysis results in a better understanding of the market, purchasing trends, budgets and help in crucial decision-making related to business [4]. Streaming analytics can help businesses identify premium clients based on their location or purchase history and will help them to reach out to the right customer at the right time. The real-time customer support also increases the confidence of the clients in the business enterprise.

16.2 Healthcare Data Sources and Basic Analytics

Healthcare data is very sensitive. It consists of medical data of patients that needs to be highly confidential. The medical researchers work on medical data that has been granted to them following confidentiality agreements and consent of medical practitioners. Using healthcare data for research is a challenge for researchers as it contains sensitive material about patients and access to that data is governed by confidentiality laws. Access to such sensitive information needs to be supported by privacy protocols and agreements. Such information is both clinical as well as non-clinical and is stored in different formats. Healthcare databases collect real-time medical data from sensors and other monitoring devices. Medical genomics has gained popularity among researchers as it deals with personalized individual medicine for individual patients. The patient data is usually unstructured in nature and mainly contains a number of medical imaging reports, such as CT, MRI, or PET scabsc.

16.2.1 Patient Data

Electronic health records (EHRs) are a computerized interpretation of a patient's medical history stored on a computer. It includes a comprehensive variety of information relevant to a case's treatment, such as demographics, issues, medications, physician's observations, vital signs, medical records, lab results, radiological reports, patient records, and billing information. Many EHRs go beyond a patients' health or treatment history to provide new insights into their care. EHRs allow physicians and associations to communicate information efficiently and effectively. In this context, EHRs are built from the ground up to be in real time, and they may be input and changed immediately by authorized practitioners. It is quite beneficial in a variety of practical situations. As an example, a health center or specialists may seek to incorporate the health records of the main practitioner into their system of record. An EHR system simplifies the process by enabling immediate access to the simplified information in real time [5]. It can assist other care-related elements such as substantiation-based decision assistance, standard operation, and problem reporting. EHRs improve the efficiency with which wellness data is stored and recovered. Patient care is improved by increasing patient engagement in the treatment process. It also helps to improve the accuracy of judgments and health conditions and the overall quality of care cooperation.

16.2.2 Medical Imaging Records

Creating and maintaining medical imaging records are crucial in the healthcare industry. These images are multidimensional and must be of high quality for clinical analysis and research purpose. These decisions can aid the treatments and prediction of new diseases based on the symptoms and response to prescribed medication. The most popular imaging modalities used to acquire a biomedical image are MRI, CT,

PET, and ultrasound [6]. These images are very helpful in the examination of the patients' internal body organs without any impact on their health condition. The medical practitioners get a better understanding of the cause of the ailment and response of the patient's body to the prescribed treatment. The primary purpose of these imaging method is to collect real-time information and assist the physician to treat their patients in critical situations, which sometimes can be life-threatening.

However, there are numerous challenges in capturing and transmitting these critical images, such as complex, diverse, and noisy images being included in this category. Object identification, categorization, recognition, and extraction of features are broad areas of image analysis research challenges. These concerns may provide crucial analytical measures used in other elements of healthcare information analytics. Once all of these issues are handled, and the development of relevant analytic measures used as input to other aspects of healthcare data analytics will be possible.

16.2.3 Sensing Device Data

A sensing device is used in the healthcare profession to detect the organs to diagnose the patient's ailment with the assistance of the signal generated by the sensor. Some medical data-gathering tools, such as the ECG and EEG, are sensors that gather impulses from different regions of the humanoid anatomy. While the fast development of sensor data has great potential for improving healthcare, it also poses a substantial issue due to data oversaturation [7]. Because of this, it is critical to build unique data analytics tools that can handle vast amounts of acquired data and turn them into valuable and interpretable information. Patients' sensor readings will be better observed, and situational awareness will be more readily available at the bedside due to such analytical tools. They will also give greater visibility into healthcare system inadequacies, which may be the source of rising healthcare expenditures.

16.2.4 Mining Clinical Notes

Medical records, such as clinical notes, store most patient information. It is often saved in an unformed data structure, and they serve as the foundation for most of the information collected in the healthcare sector. Physicians may enter clinical information directly into these databases or use voice recognition programs to record dictation. These are, without a doubt, the most significant reservoir of untapped knowledge available. To say that hand-coding in free-text form of a wide range of clinical data is prohibitively expensive and time-intensive is an understatement, even if confined to primary and secondary diagnoses, as well as processes for billing reasons. The difficulty in automatically analysing such notes is in translating clinical material accessible in a free-text format into a structured one. Their unstructured nature, variability, multiple forms, and shifting context across different patients and practitioners make it difficult to interpret them effectively.

A significant role is played by natural language processing (NLP) and entity extraction in inferring meaningful knowledge from vast quantities of clinical data and automating the encoding of clinical information promptly [8]. Data preparation procedures are more critical than mining algorithms alone in many cases. Processing clinical text using NLP approaches is more difficult when compared to processing another type of text because of the ungrammatical character, telegraphic sentences, dialogues, shorthand vocabulary and often misspelt clinical terminology. A broad variety of NLP and data mining approaches are used to extract information from clinical data.

16.2.5 Community Data Analysis

The rapid rise of numerous community tools, such as social media platforms, blog sites, forums, automatic speech recognition facilities, and virtual groups, offers a plethora of data regarding communal perception on several healthcare-related topics. Multiple patterns and data may be extracted from online data, which can then draw beneficial conclusions regarding demographic health and health supervision [9]. It is possible to greatly minimize the delay associated with acquiring such complicated information by effectively evaluating these enormous amounts of knowledge. Social media information often reflects disease epidemic detection, and examining the content's historical context may give useful insights about disease outbreaks. A significant announcement regarding various medical disorders is found on the internet. Medical illnesses are common among different people. Due to the data's unreliability, social media analysis presents a considerable difficulty since the findings must be evaluated with care.

16.3 Retail Domain

16.3.1 Retail Industry

Retailing is one of the most successful and diverse industries in the world. The fact is that there are various sorts of retail outlets, such as supermarkets, pharmaceutical stores, department stores, beautician stores, and so on, and there is neck-to-neck war among all retailers, which strengthens the increasing rivalry among all of them. This rivalry takes the shape of same-store kinds in terms of price reductions and other incentives to entice shoppers to buy. Currently, commerce has transformed disordered stacks into nicely organized storefronts [10].

However, the fundamental difficulty for today's merchant is that the contemporary client is concerned with quality and brand. They can evaluate the services supplied by numerous stores from the comfort of their own homes with a simple click on a computer or smartphone. As a result, clients choose to buy from e-commerce websites rather than physically visiting a retail shop, resulting in a decrease in sales for merchants, who are now facing a major challenge from online competitors. The result is that retail establishments are obliged to strive diligently to meet and exceed their customers' expectations by bringing all of their desired items together in one location. As a result, the key goal for today's merchants is to maintain their existing client base while avoiding being swayed by intense competition.

Streaming analytics may assist in the transformation of business insights by obtaining data from within the organization, such as enterprise resource planning (ERP) and investment management, and from other sources, such as network edge and public assets, connecting it the ultimate decision. From acquiring historical information to the most recent data, streaming analytics may assist merchants in navigating existing issues and anticipating any upcoming problems.

16.3.2 Customer Satisfaction in Real-Time Analysis

Various bits of research have been carried out on consumer satisfaction across the retail industry and its factors. It is not possible to make broad generalizations about effectiveness throughout all retail sectors and kinds of distribution. The quality requirements are evaluated differently but usually contain the following facets: pricing, availability of products, and guidance. With the help of mobile devices and the Internet of Things (IoT), merchants can engage with their consumers more personally. These technological advances may be used as a data source to obtain information about client sentiment. The ability to personally identify clients and supply them with information on pricing and services is made simpler for shops [11]. They can take immediate action to meet client demand and maintain a competitive advantage. They will be able to predict what the consumers want next since they will have access to real-time client data. They will be able to accommodate their demands and provide newer items or services early.

Real-time analytics systems can identify and handle any issues that may develop as soon as possible, preventing them from harming the company and ensuring that consumers always receive excellent care imaginable. It provides tangible benefits, assisting businesses in attracting new consumers and increasing revenue by developing customized experiences based on rich, immediate location- and behavior-based data. In recent years, Big Data analytics has grown to incorporate batch processing, online processing, streaming, and offline data processing, which was formerly the primary mode of operation. In general, it is hard to exert control over online shopping networks; as a result, the ideal approach is to use the networks for one's own gain.

16.3.3 Technological Applications in the Retail Domain

According to several retail industry specialists, the latest technological advances enable merchants to remain competitive in four important categories: (1) accessibility; (2) pricing; (3) size; and (4) velocity. Technology enables strategy and selection at the management level. Various data-mining tools aid in inventory management, pricing, marketing choices, and product creation and design.

16.3.3.1 Inventory Tracking

- *Electronic data interchange (EDI)*: Computer-to-computer interactions from the shop to the suppliers' databases and online ordering.
- *Hand-held inventory devices used wirelessly*: Maintain an inventory and upload the information to the database at headquarters.
- *Universal product code (UPC)*: Products are identified and tracked via the use of barcodes and unique numbers.
- *Automatic replenishment*: Handles the refilling of items that have been sold.
- *Virtual shelves*: Intranets between merchants and providers allow faster communication and more accurate inventory management.

16.3.3.2 Customer Service

- *CRM software*: Shoppers' information may be gathered by merchants using this tool.
- *CD-ROMs at the register*: Allow sales representatives to place specific orders at the moment. In addition, sales training is provided to sales workers on the floor.
- *Kiosks*: Deliver product information to your consumers.
- *Point-Of-Sale (POS) terminals*: Coupons and reports may be printed, frequent buyer discounts can be calculated, buyer identity information is captured, labor hours can be scheduled, and email connections between stores and headquarters are also used.
- *Fingerprint technology*: Input for credit card purchases at the point of sale terminal. A digital copy of the receipt is preserved.
- *E-commerce technology*: It enables companies and consumers to interact anywhere throughout the world.

16.3.3.3 Data Warehousing

- *Executive Information Systems (EIS)*: Create charts of detailed data to aid retail professionals in business decisions in their various sectors by analysing and visualizing the data.

16.3.4 Retail Data Sources and Analysis

The sources of consumer, sales and operational data available to retailers have expanded significantly during the previous decade. From more conventional to more creative sources, they all play a critical role in ensuring that they obtain the information necessary to provide correct and full sales analytics:

- *POS data*: Systematically tracking all sales transactions, commonly known as point of sale (POS) systems, generates massive amounts of sales information, generating patterns, movements, and anomalies.
- *e-commerce transactions-related data*: The website backend systems record all online sales activity in logs and events archived. When there is a need to collect data streaming over an API or an XML connection, it is necessary to use cutting-edge technologies. Data may conceal valuable information about searches and inventory requests, as well as information about fields that are not being supplied.
- *In-store sensor data*: Almost all stores now have Wi-Fi technology installed, and the larger ones have sensors and Internet-of-Things devices installed. We can get incredibly useful information from all these, such as client activities throughout the shop, foot traffic within the store, and so on.
- *Supply chain systems*: Get the benefit from predictive analytics for inventory management by using data supplied by supply chain systems.
- *Social media*: Social media accounts and customer reviews are valuable sources of information, especially when it comes to knowing customer requirements and perceptions and following competitor activity and strategies.

Data analysis using Big Data has been hailed as a revolutionary technology that can transform the retail business. Despite the fact that data management techniques have several advantages, an increasing number of firms are using data analysis through large datasets to get information into their processes and boost their revenue production capabilities [12].

16.4 Different Ways of Streaming Analytics

The streaming data is analysed in different ways using streaming analysis tools. Making stream processors capable of quick computing and modern work with many data streams requires the development of dedicated technology. These technologies are essential in the development of a streaming analytics platform. Table 16.1 shows the main platforms used in streaming analytics.

16.5 Real-Time Streaming Analytics in Healthcare Use Cases

Doctors' and nurses' workloads have been greatly reduced thanks to the rise of real-time streaming analytics technologies in the recent decade. Streaming analytics and

Table 16.1   Main Platforms Used in Streaming Analytics

Platform	Streaming analytics projects	Description
Apache	Apex	Apex is a Hadoop platform powered by YARN. It can handle unbounded data sets and integrate with various data portals. Apex provides high efficiency, reduced latency, and unified structures.
	Flink	Apache Flink is a decentralized streaming process system built in Scala and Java that is available as an open-source project. The Flink runtime is capable of supporting group and stream computing, also sequential algorithm.
	Samza	Samza is a computing framework that enables blunder, persistent, and efficient domain-specific stream computing via straightforward API. This application communicates with other applications over Kafka and operates under YARN.
	Spark	Spark is an open-source, general-purpose cluster computing platform that does analytics, ETL, machine learning, and graph analysis on data in motion or repose. It is available as a free download from the Apache Software Foundation. Spark has indeed been proved in production for a wide range of use cases, and it conveniently supports SQL querying.
	Strom	Storm offers very low latency, making it ideal for applications that need near real-time processing. As opposed to most other systems, it can process enormous amounts of data and deliver conclusions with far reduced latency.
Amazon	Kinesis	Real-time streaming services provided by Amazon Kinesis are durable and scalable. In a single second, it may gather terabytes of data from hundreds of numerous sources, such as databases and financial activity.

(Continued)

Table 16.1 (Continued)

Platform	Streaming analytics projects	Description
	Firehose	Amazon Firehose makes it possible to link data streams into current business intelligence tools, analytical edges, and warehouses. It assists in retrieving data and integrating it into current data warehousing systems.
Azure	Azure Stream Analytics	It works with Power BI to analyse data. Both systems are cloud-managed, and it combines data from numerous sources and deliver low-latency processing.
	Power BI	Power BI is a public analysis tool. It is possible to integrate Stream Analytics with Power BI to update dashboards and create visualizations using interactive components and settings.
Google Cloud	Google Cloud Stream Analytics	A specialized engine for data input, computing, and analysis is built into it, it offers capabilities equivalent to Amazon Simple Storage Service.
Oracle	Oracle Stream Analytics	Its cloud-based platform provides stream ingestion, processing, and visualization tasks in a single environment.
IBM	IBM Streaming Analytics	IBM Streaming Analytics allows you to construct real-time analytics apps. Ingestion/ transformation/analysis of data using IBM Streams. Streams may be implemented on-premise or on the IBM Cloud.

Internet-of-Things technologies are being used by top healthcare providers to identify trends and patterns more quickly. Consequently, patient care and expenses are
both being improved. Healthcare providers may use predictive analytics on data in
motion for continuous choices, allowing them to record and analyse data all the
time, just in time, with streaming analytics. The ultimate goal is to save lives, reduce
hospitalizations, and improve the health of communities through preventive care.

16.5.1 Data Stream Computing in Healthcare

16.5.1.1 Stream Computing Technology: Apache

Apache is one of the most important stream processing technologies. Kafka is a scalable, fast publish-subscribe messaging system, and long-lasting. Thousands of consumers sending several megabytes of requests and writes per second can be handled
by a Kafka broker. The Kafka messaging system facilitates a standard healthcare data
collection as well as analysis, as seen in Figure 16.1.

A Kafka topic is where a producer can post messages (messaging queue). A subject is another word for topic. In addition to a message feed category, an inter-
mediary for Kafka, also known as a Kafka host, creates new topics. Kafka brokers
can save information for later use. Consumers subscribe to Kafka topics, which then
extract messages from the brokers. Standard data warehouses store data for online
analysis, while offline users ingest messages and store them in Hadoop or offline for
analysis, if there is a data warehouse. These customers have the ability to use any

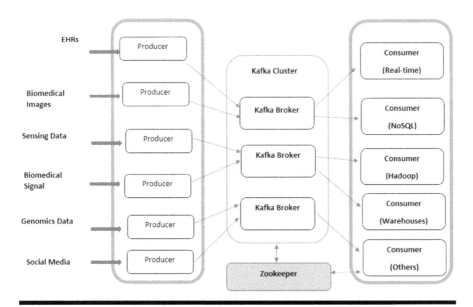

Figure 16.1   Messaging system.

NoSQL database or a memory filter to store results, send out alerts to the appropriate groups, and ingest messages. To keep track of message offsets, providers and customers use ZooKeeper [13].

16.5.1.2 Healthcare Analytics with Big Data: A Generic Architecture

Stream and batch computing are combined in this area to improve Big Data processing in healthcare. With low latency and huge datasets, it can also reduce costs and improve healthcare conditions, as shown in Figure 16.2. It is possible to separate stream computing, which uses Kafka, Storm, and NoSQL Cassandra for real-time computing, database computing and batch computing, which uses Hadoop clusters and HBase databases to store output data. At the serving layer, a query management interface will emerge between real-time and batch views of data.

Kaka producers generate many message queues on a regular basis in the stream computing layer. At Kafka brokers, topics are subdivided and sent downstream to Storm clusters. This information can be obtained through electronic health records, sensors, and social media. Because the data is computed, the outcome is kept in Cassandra, a NoSQL database, due to its partition-tolerance and easy availability features in systems with several nodes. Delta findings are displayed in real time to improve batch data. The batch layer, which manages HDFS's immutable

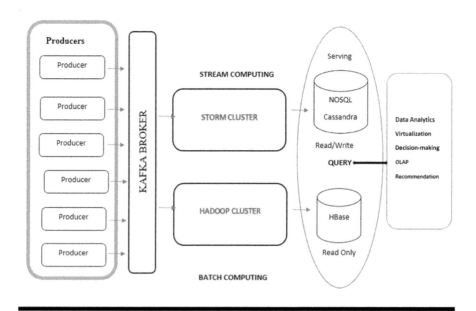

Figure 16.2 Healthcare system architecture.

append-only massive data set, is primarily in charge of pre-computing views, based on MapReduce queries. The computation of the view is a never-ending process. When a new data item is received, the views will be updated. The view must be generated from the entire dataset because the batch layer will not be updating the view on a regular basis. Computation time grows in proportion to the size of the dataset and the number of nodes. In the domain of Big Data, the HBase and Hadoop frameworks are useful if you have a lot of data [14, 15]. The serving layer indexes and exposes views to the querying system, thus they can be questioned. A set of files with pre-calculated perspectives is created by the batch layer. The serving layer just needs to offer batch modifications and random requests because batch views are fixed. By combining batch and real-time views, any query can be answered.

16.6 Medical Signal Analysis

Physiological signal monitoring and telemetry equipment can be found almost anywhere. A short-term loss of these monitors' collected data makes in-depth analysis of previously collected data difficult. Recent studies [16] have shown that telematics patient care and management are improved by continuous physiological time series monitoring. More investigation, however, will be required. The systematic application of a steady waveform (a moment signal) and associated medical record data to support standardized and scalable decision-making for patient care, as defined by applied analytical disciplines, is referred to as "streaming data analytics" in healthcare (for example, statistical, quantitative, contextual, cognitive, and predictive analytics). The following steps can be taken to analyse streaming waveforms in clinical situations:

- *Data collection and ingestion*: This can begin with the use of a streaming collection and ingestion platform capable of handling a wide range of waveforms with varying degrees of precision, accuracy, and reliability. Dynamic waveform data must be integrated into the electronic health record (EHR). Analytics engines can gain situational and contextual awareness by analysing EHR data. By increasing the amount of data used, it is important to maintain a healthy balance between predictive analytics' sensitivity and specificity. The study's illness cohort will include a significant impact on signal processing. A large number of target features can be extracted using signal processing algorithms, they are then analysed by a machine learning model that has been pre-trained to produce real-time actionable insights. These breakthroughs could pave the way for new diagnostics, predictions, and prescriptions. In response to the data, additional procedures, such as alerts and notifications to doctors, could be developed. Combining discrete data from various sources with consistent waveform data is a critical step in developing new diagnostics and therapies. Before these systems can be used in clinical settings, a number of system, analytic, and clinical needs must be addressed. When designing monitoring systems that can accept both continuous and discrete data from non-continuous sources, there are a number of challenges and solutions to consider.

■ *Data analysis and interpretation*: Prior to this study, physiological signal data collected throughout time was seldom streamed in real time or saved for an extended period of time. Data captures were frequently time-limited and could only be recovered if the device-makers provided proprietary software and data formats. Regardless of whether the majority of major medical device manufacturers have begun to develop interfaces for obtaining data in motion, such as live streaming data from their equipment, traditional Big Data challenges are due to its high rate of movement. In the current situation, governing issues such as data protocols, data standards, and worries about data privacy are all factors. Another impediment to adoption by the masses of real-time data collecting has been a lack of network bandwidth, scalability, and cost [17]. Medical research communities-specific programs [18] can now be implemented system-wide [19]. The scientific community is considering continuous monitoring systems based on data generated by live monitors [20]. Several indigenous and off-the-shelf attempts have been made to design and install data collection systems [21]. Thanks to new industry technologies, data collection from patient monitors is becoming easier across a wide range of healthcare systems, regardless of the device vendor.

■ *Data retrieval and storage*: To handle the huge amount of data on the cloud and other patient information that could come from a therapeutic setting, a dependable method of data storage is required. Given how computationally and time-intensive storing and retrieving data can be, it's vital to have a storage facility architecture that enables speedy information to be pulled and saved in response to analytic needs. Because of their ability to keep and process huge amounts of information, healthcare researchers are increasingly turning to computer systems: Hadoop, MapReduce, and MongoDB are examples of such systems [14]. Traditional relational databases can be replaced by document-oriented databases that run on a variety of platforms using MongoDB. Interoperable health systems frequently have their own relational database schemas and data models, making it difficult to share healthcare data across institutions or conduct research projects. To make matters worse, traditional databases do not support the integration of data from multiple sources (e.g., streaming waveforms and EHR data). When it comes to healthcare information, document-based databases like MongoDB can help with high speed, high availability, and easy scalability. Open-source frameworks such as Apache Hadoop can now be used to distribute the processing of massive data sets across a distributed network of computers. MapReduce and Spark are just two of the numerous computing modules that can run on this massively scalable platform [14]. Because it can consume and process streaming data and provides machine learning and graphing tools, an analytics module like Spark is ideal for continuous telemetry waveforms. The ultimate goal of using such tools for researchers is to quickly and efficiently translate scientific discoveries into medicinal applications, both in real time and in retrospect.

■ *Data aggregation*: This is the process of merging data from several different sources. Integrating data from multiple sources and maintaining consistency are just two of the challenges that healthcare data aggregation faces. Aside from data quality and standardization, two other issues must be addressed: Because medical data is frequently complex, interconnected, and interdependent, it is critical to keep it as simple as possible. Secure medical data storage, access, and use are critical to the privacy and provenance of medical records [22]. The analysis of continuous data makes considerable use of time-related information. The dynamic nature of the time context during integration may greatly increase the technical challenges when combining measured data with fixed electronic health record data, because static data does not necessarily provide real-time context. It takes time and effort to create an open database of waveforms and other electronic medical data that is accessible to researchers all over the world [23]. MIMIC II [24] and other datasets contain signals and other medical data from various real-life populations of patients.

16.7 Big Data Analytical Signal Processing

Clinical decision support systems (CDSSs) based on Big Data are becoming increasingly popular in signal processing research [25]. For many years, organizations, such as the Institution of Medicine, have strongly pushed health information technology (including CDSS) [26]. Clinical decision support systems provide doctors with patient-specific data and knowledge that has been intelligently vetted and delivered at the right time to assist them in providing better care.

In intensive care units (ICUs), each patient generates a significant amount of physiological data, which is quickly accumulated. Researchers discovered an increased risk of developing CDSS in ICUs. Using information gathered from severely ill individuals in the intensive care unit, an architecture for building a management system for caring for a patient has been provided [27]. This infrastructure, which includes fix and streaming data from ICU patients who are seriously unwell, can be used for data mining and real-time alerts of critical events. Streaming data from various sources, clinical decision support devices, such as infusion pumps, EEG monitors, and cerebral oxygenation monitors, can be used in a neonatal intensive care unit [28]. In a current clinical trial, biomarkers derived from real-time successful estuation can be predicted via signal processing of cardiac and breathing signals [29]. An animal study established the use of non-invasive time series such as oxygen delivery, water content, and blood circulation as indicators of soft tissue healing in wound care.

Electrocardiographic data collected via telemetry was coupled with demographic characteristics to develop an in-hospital cardiac arrest early detection system based on medical record, ejection fraction, lab results, and medicines [30].

The MIMIC II database was used by Lee and Mark [31] to use cardiac and blood pressure time series data to prompt therapeutic treatment in hypotensive episodes. Another study [32] found that physiological waveform data could be combined with clinical data from the MIMIC II database and be used to detect shared characteristics among patients in specific cohorts. This similarity may improve caregiver decision-making because caregivers can draw on information gained from similar illness states to make informed decisions. To identify patients with cardiovascular instability, a combination of waveform data from the MIMIC II database is used [33]. Pre-operative care and operating room settings can collect a wide range of physiological data that can be analysed in [34] to track the patient's condition during and after the procedure. [35] demonstrated that machine learning models based on data fusion can improve diagnosis accuracy when using breath omics biomarkers (in an inhaled air metabolomics study) as a diagnostic aid.

Neurology researchers have recently become interested in the use of electrophysiologic monitoring in patient treatment. This monitoring could be used in the near future to develop novel diagnostic and treatment procedures, as well as shed new light on difficult illnesses. According to an article [36], many physiological monitoring devices have been developed specifically for the treatment of patients in need of neurocritical care. Due to the efforts of [37], patients with severe brain injuries can now benefit from multimodal monitoring and care that is tailored to their specific needs. Researchers used multimodal brain monitoring to see if patients' reliance on mechanical ventilation and hospitalization could be reduced. [38] discusses the fundamentals of multimodal monitoring in neurocritical care for secondary brain damage, as well as the first and future approaches to understanding and applying such data to clinical care.

Patients and enthusiasts of all ages are using personal monitoring devices, which are becoming more affordable, portable, and easy to use outside of professional settings. However, combining data from multiple portable devices at the same time, as with clinical applications, could be difficult. Pantelopoulos and Bourbakis [39] examined wearable biosensor research and development. They discussed the advantages and disadvantages of working in this industry. A network-based tele-medicine study [40] employs the same portable blood pressure and body weight equipment. According to [41, 42], a variety of stationary and mobile sensors that can be used to improve patient care technology can help with healthcare data mining.

16.8 Big Data Analytics in Genomics

In the United States, Big Data analytics in the healthcare sector plays a significant role in saving billions of dollars. Big Data will assist in forecasting and preparing responsive actions to a disease pandemic, increasing the efficiency of clinical trial tracking, and maximizing healthcare spending at various levels, from patients to the hospital premises, then to the government. Genomic sequencing, which is predicted to be the future of healthcare, is another significant field. This section examines the opportunities, ongoing research, and challenges of genomics in the context of emerging Big Data and analytics. Genomic medicine intends to construct individualized diagnostic strategies or efficient rehabilitation decision-making by using patient's genomic details.

Big Data analytics explores large-scale data sets to retrieve relevant patterns, hidden correlations, and other insights. While integrating and manipulating wide-ranging EHRs and complex genomic details on a Big Data system presents several challenging issues, it also offers a reasonable strategy to establish an effective methodology for identifying the clinical genetic variants in order to proceed with appropriate diagnosis and therapy. The complexities of processing Next-Generation Sequencing (NGS) data on a large scale and complex clinical data obtained from EHRs for genomic medicine are discussed. In order to implement this, the potential solutions are presented to handle various challenging issues in manipulating, handling, and evaluating clinical and genomic data. A realistic Big Data toolkit is introduced for detecting active genetic variants using NGS and EHR data.

A few scenarios are discussed here, illustrating how the healthcare industry can use Big Data analytics in a variety of ways. Personalized medicine is a hot topic among different services offered by healthcare organizations, and medicines are now easily available for the general public. These drugs perfectly function to detect particular diseases. This is due to the individuality of each person, as well as the interdependencies and dependencies that every individual personal characteristic possesses. Moreover, personalized medicines that exploit particular details of patients such as genomics data could be generated in the future with the aid of Big Data, focused on preparing similar kinds of patient profiles and their responsive actions to those strategies.

Genomics is widely known as the analysis of an organism's entire genetic material. Deoxyribonucleic acid (DNA) in the form of a double helix is actually the inherited component. The nucleotides such as Adenine (A), Cytosine (C), Guanine (G), Thymine (T) are the four distinct forms of chemical bases that make up DNA. The double helix is constructed by twisting pairs of A, C, G, T into a ladder form. Furthermore, among pairs, A and T are exclusively blended with each other, while G and C do the same. The way nucleotides are organized determines the uniqueness of each human; each spring of DNA is distributed in various orders and has diverse lengths. Though the enzymes are highly responsible for DNA transcription and translation, sometimes errors are introduced, resulting in even more fatal mutations or changes from the standard genome code. And, changing a single base in nucleotides from "ACGTTTGA" to "ACGTCTGA" may have negative health implications.

Interpreting genetic data brings us closer to medicine's future; it signifies the start of an age characterized by personalized medicine, in which every human has their own personal DNA sequence and, thus, their own personal genome, loaded with numerous mutations, in that some are beneficial and others dangerous. The Human Genome Project (HGP), which is sponsored by the US government, completed the first trial at sequencing in 2003. From 1990 to 2003, HGP received $3 billion in funding to convert sequences from a manual form to an automated process. Moreover, the HGP has discovered that in total there are over 20,500 human genes. And, the human genome was determined in three ways by scientists working on HGP,

1. Determine the "order"/"series" of the complete genome bases in human DNA.
2. Generate maps which display the positions of genes for large parts of the chromosomes.
3. Produce connectivity maps, and difficult versions of its type first developed in early Drosophila research, with this, the inherited properties can be monitored for several generations.

Genome Wide Association Studies (GWAS) move the Human Genome Project a step forward by identifying the most common genetic variants in various people and establishing links among genes and diseases. Some challenges in the manipulation of genomic data are:

■ Despite the fact that around 6000 Mendelian disorders are examined at the primary genetic level, the majority of their activities in health and disease remain unknown.
■ The National Institutes of Health (NIH), Sequence Read Archive (SRA) databases have developed at an exponential pace over the last eight years. While advances in NGS have made it simple to sequence an entire genome/exome, the handling, analysis, and interpretation of the genomic data produced by NGS remain a significant challenge.

By considering the human genomes for about 3 billion base pairs, sequencing a human genome creates about 100 terabytes of data in the file format of BAM (Binary Version of Sequence Alignment/Map), and VCF (Variant Call Format). Moreover, the BAM file size in a sequencing experiment is found by the coverage and read length. For a single sample of 30 WGS data, the FASTQ file could be around 250 GB, the BAM file could be around 100 GB, the VCF file could be around 1 GB, and the other annotated files could be around 1 GB. And, by considering the estimated file sizes of various NGS data formats, it is time-consuming to generate such files.

Data can be analysed swiftly with the Big Data infrastructures. In comparison to the original version, a Big Data-based Burrows-Wheeler Aligner (BWA) will maximize the alignment speed 36-fold.

Currently, most sequencing data analysis methods depend on VCF files, which believe that all "no-call locations" are just like reference alleles. In reality, inadequate coverage can be the cause of several "no-call locations." As a result, data quality information for each location, which includes coverage and Phred-scaled base quality ratings, must be used to determine if "no-call locations" are reference-consistent owing to excessive coverage or reference-inconsistent due to the less coverage in the downstream data review. Exome sequencing data has been examined using a variety of toolkits for cloud computing, data compression, to prioritize variants, data exchange, Copy Number Variation (CNV) identification, and phenotypes. Analytical tools work on VCFs, which do not generally require an infrastructure always since VCFs are quite a bit smaller than BAM files.

Researchers are now encountering considerable difficulties in maintaining, processing, manipulating, interpreting, and analysing WGS data. Thus, the problems will be compounded as millions of people are sequenced, that becomes one of the targets of the Precision Medicine Initiative (PMI) in the United States and related work around the world. It may be possible to build a Big Data framework to handle and interpret large volumes of genomic data that is consistent with clinical workflows by exploiting the scalability and distribution features in Big Data infrastructures.

16.8.1 Securing Genomic Data

Genomic data must be secured. Hence, confidentiality and privacy should be preserved in the same way as securing other health information. Secured data transmission, password protection, data encryption, reviewing data transfer strategies, and the operational strategies against breaching activities are required. The Fair Information Practices Principles (FIPPs) introduced a paradigm that allows data sharing based on Health and Human Services' guidelines followed by the US Department of Health. Some of its principles are individual data accessibility, data collection, data transparency, disclosure constraints, data integrity, data quality, data accountability and data protection. The Workgroup for Electronic Data Interchange (WEDI) has published a paper highlighting the problems of health IT integration in terms of infrastructure, procedures, and coordination [43]. Data integration, data access, data exchange, and data governance are the challenges. Advances in cloud computing technologies make it easier to store big genetic data files and combine data to make it more accessible.

A study has been presented which outlines the complexities of health IT integration in terms of coordination, workflows, and infrastructure. Some more challenges are data convergence, data sharing, and data governance. Since data storage services are offered by an outside agency of the healthcare institution, cloud computing poses potential security issues. Cloud providers are perceived as business associates and they must provide consent on a business associate agreement to comply with the HIPAA privacy, protection, and violation of notification laws. Managing data access and the implementation of a user role-based access system are the two ways that can be followed by cloud service providers in order to resolve these concerns.

Some issues that arise while handling genomic data are described below,

16.8.2 Privacy

Privacy is one of the most sensitive issues in data science. In fact, the public's current privacy concerns regarding surveillance cameras, financial transactions, and email communication are crucial nowadays. Privacy disclosures are commonly difficult to detect due to the cross-reference of larger datasets, quasi-identifier is the best example for this. While genomics-based privacy has some distinct characteristics when compared to data science-based privacy issues, the fact is that the genome is crossing across generations and is very important to the public. Moreover, revealing genomic data is more dangerous than leaking other kinds of data. Once a person's data or their related variants have been revealed or leaked, they may not be able to take them back. Finally, genomic data are far larger in size than several other kinds of personal data; for example, the genome contains much more confidential personal information than a social security number or credit card.

16.8.3 Data Ownership

Privacy is a major concern in data management. An individual entity or patient owns their personal information, a coercive trend in this biomedical science is that the researcher who produces a dataset manages it. Researchers with large datasets will have a long history that supports them in analysing the data over the period in order to discover interesting stories. In particular, health data has commercial and medical value, so corporations and governments often claim control and ownership of large datasets. According to data miners, all information should be accessible, since this would allow easy aggregation of massive amounts of data, better statistical analysis, and optimal outcomes. In fact, combining larger datasets results in increasingly better genotypes being correlated with phenotypes. Furthermore, we estimate that sharing biases, such as particular diseases, pathologies, and phenotypes, and being more enthusiastic to share genetic data would occur in an ideal case, in which individual users were committed to free access and the dataset became entirely accessible and freely transferred by the users.

Skew in the dataset can be caused by education, socio-economic status, and healthcare access, all of which can bias mining efforts, including knowledge extraction. Likewise, around 80 percent of participants in genome-wide association studies are of European ancestry, despite the fact that group makes up only 16 percent of the global population. As a result, complete data sharing is unlikely to be practical in the future for the best genomic studies. The development of a large private archive may be one potential technological solution for exchanging genomics data. This is completely distinct from the World Wide Web, and which is essentially a public resource. To allow data sharing and suggest a way to centralize the computational storage of larger datasets to maximize performance, a private archive will be approved for use only by certified researchers. Furthermore, in future, people will be faced with challenging data science problems, namely sharing constrained data in a certain context. Finally, data ownership is closely associated in obtaining profits and benefits from the data. The best strategies have to be identified not only to recognize the data generation process, but also to analyse the data efficiently and to value the reward analysts and data generators.

16.9 Conclusion

This chapter discusses real-time streaming analytics in the healthcare and retail domain. Developing data analytics is now easier than ever before, thanks to the digital transformation of business. Data streaming analytics may enhance decision-making by providing better business insight predictions. Stream analytics' ultimate goal is to assist organizations to stay ahead of their competition by allowing them to make smarter decisions. Streams of data are gathered in retail and healthcare via different electronic devices such as mobile phones, wireless sensors, etc., linked through the internet. The healthcare business requires software tools and technology capable of real-time analysis of massive and diverse data streams. The streaming analysis technology is used in retail establishments and e-commerce because increasing client happiness is the key aim. Streaming analytics also helps new companies reach their target audiences and gain market share. Big Data analytics uses both structured and unstructured data. It enhances decision-making for both researchers and medical practitioners.

References

1. Kantardzic, M., and Zurada, J. (2016) "Introduction to Streaming Data Analytics and Applications Minitrack." In *2016 49th Hawaii International Conference on System Sciences (HICSS)*, IEEE, January, pp. 1729–1729.
2. Manyika, J., Chui, M., Brown, B., Bughin, J., Dobbs, R., Roxburgh, C., and Hung Byers, A. (2011) *Big Data: The Next Frontier for Innovation, Competition, and Productivity*. New York: McKinsey Global Institute.
3. Oyelade, J., Soyemi, J., Isewon, I., and Obembe, O. (2015) "Bioinformatics, Healthcare Informatics and Analytics: An Imperative for Improved Healthcare System." *International Journal of Applied Information System*, 13(5): 1–6.
4. Yadav, S. A., Sharma, S., Das, L., Gupta, S., and Vashisht, S. (2021) "An Effective IoT Empowered Real-time Gas Detection System for Wireless Sensor Networks." In

2021 International Conference on Innovative Practices in Technology and Management (ICIPTM), IEEE, February, pp. 44–49.

5. Singh, A., Sharma, S., Kumar, S. R., and Yadav, S. A. (2016) "Overview of PaaS and SaaS and its Application in Cloud Computing." In *2016 International Conference on Innovation and Challenges in Cyber Security (ICICCS-INBUSH)*, IEEE, February, pp. 172–176.

6. Sternberg, S. R. (1983) "Biomedical Image Processing." *Computer*, 16(01): 22–34.

7. Aggarwal, C. C. (Ed.) (2013) *Managing and Mining Sensor Data*. Singapore: Springer Science and Business Media.

8. Meystre, S. M., Savova, G. K., Kipper-Schuler, K. C., and Hurdle, J. F. (2008) "Extracting Information from Textual Documents in the Electronic Health Record: A Review of Recent Research." *Yearbook of Medical Informatics*, 17(01): 128–144.

9. Sadilek, A., Kautz, H., and Silenzio, V. (2012) "Modeling Spread of Disease from Social Interactions." Paper presented at Sixth International AAAI Conference on Weblogs and Social Media, May.

10. Smith, A. D. (2008) "Modernizing Retail Grocery Business Via Knowledge Management-Based Systems." *Journal of Knowledge Management*, 12(3): 114–126.

11. Weber, F. (2019) "Streaming Analytics—Real-Time Customer Satisfaction In Brick-And-Mortar Retailing." In *Proceedings of Computer Science On-line Conference*. Cham: Springer, pp. 50–59.

12. Davenport, T. H. (2006) "Competing on Analytics." *Harvard Business Review*, 84(1): 98.

13. Hunt, P., Konar, M., Junqueira, F. P., and Reed, B. (2010). "{ZooKeeper}: Wait-free Coordination for Internet-scale Systems." Paper presented at 2010 USENIX Annual Technical Conference (USENIX ATC 10).

14. Vora, M. N. (2011) "Hadoop-HBase for Large-Scale Data." In *Proceedings of 2011 International Conference on Computer Science and Network Technology*, IEEE, vol. 1, pp. 601–605.

15. Taylor, R. C. (2010) "An Overview of the Hadoop/MapReduce/HBase Framework and Its Current Applications in Bioinformatics." *BMC Bioinformatics*, 11(12): 1–6.

16. Apiletti, D., Baralis, E., Bruno, G., and Cerquitelli, T. (2009) "Real-Time Analysis of Physiological Data to Support Medical Applications." *IEEE Transactions on Information Technology in Biomedicine*, 13(3): 313–321.

17. Menachemi, N., Chukmaitov, A., Saunders, C., and Brooks, R. G. (2008) "Hospital Quality of Care: Does Information Technology Matter? The Relationship Between Information Technology Adoption and Quality of Care." *Health Care Management Review*, 33(1): 51–59.

18. Raghupathi, W., and Raghupathi, V. (2014) "Big Data Analytics in Healthcare: Promise and Potential." *Health Information Science and Systems*, 2(1): 1–10.

19. Ahmad, S., Ramsay, T., Huebsch, L., Flanagan, S., McDiarmid, S., Batkin, I., and Seely, A. J. (2009) "Continuous Multi-Parameter Heart Rate Variability Analysis Heralds Onset of Sepsis in Adults." *PloS One*, 4(8): e6642.

20. Goldberger, A. L., Amaral, L. A., Glass, L., Hausdorff, J. M., Ivanov, P. C., Mark, R. G., and Stanley, H. E. (2000) "PhysioBank, PhysioToolkit, and PhysioNet: Components of a New Research Resource for Complex Physiologic Signals." *Circulation*, 101(23): e215–e220.

21. Kaur, K., and Rani, R. (2015) "Managing Data in Healthcare Information Systems: Many Models, One Solution." *Computer*, 48(3): 52–59.
22. Uzuner, Ö., South, B. R., Shen, S., and DuVall, S. L. (2011) "2010 i2b2/VA Challenge on Concepts, Assertions, and Relations in Clinical Text." *Journal of the American Medical Informatics Association*, 18(5): 552–556.
23. Athey, B. D., Braxenthaler, M., Haas, M., and Guo, Y. (2013) "tranSMART: an Open Source and Community-Driven Informatics and Data Sharing Platform for Clinical and Translational Research." Paper presented at AMIA Summits on Translational Science Proceedings, 2013, 6.
24. Scott, D. J., Lee, J., Silva, I., Park, S., Moody, G. B., Celi, L. A., and Mark, R. G. (2013) "Accessing the Public MIMIC-II Intensive Care Relational Database for Clinical Research." *BMC Medical Informatics and Decision Making*, 13(1): 1–7.
25. Belle, A., Kon, M. A., and Najarian, K. (2013) "Biomedical Informatics for Computer-Aided Decision Support Systems: A Survey." *The Scientific World Journal*, 2013. doi:10.1155/2013/769639
26. Bloom, B. S. (2002) "Crossing the Quality Chasm: A New Health System for the 21st Century (Committee on Quality of Health Care in America, Institute of Medicine)." *JAMA, Journal of the American Medical Association-International Edition*, 287(5): 645.
27. Han, H., Ryoo, H. C., and Patrick, H. (2006) "An Infrastructure of Stream Data Mining, Fusion and Management for Monitored Patients." Paper presented at 19th IEEE Symposium on Computer-Based Medical Systems (CBMS'06), IEEE, June, pp. 461–468.
28. Bressan, N., James, A., and McGregor, C. (2012) "Trends and Opportunities for Integrated Real Time Neonatal Clinical Decision Support." In *Proceedings of 2012 IEEE-EMBS International Conference on Biomedical and Health Informatics*, IEEE, pp. 687–690.
29. Seely, A. J., Bravi, A., Herry, C., Green, G., Longtin, A., Ramsay, T., ... and Marshall, J. (2014) "Do Heart and Respiratory Rate Variability Improve Prediction of Extubation Outcomes in Critically Ill Patients?." *Critical Care*, 18(2): 1–12.
30. Attin, M., Feld, G., Lemus, H., Najarian, K., Shandilya, S., Wang, L., ... and Lin, C. D. (2015) "Electrocardiogram Characteristics Prior to In-Hospital Cardiac Arrest." *Journal of Clinical Monitoring and Computing*, 29(3): 385–392.
31. Lee, J., and Mark, R. G. (2010) "A Hypotensive Episode Predictor for Intensive Care Based on Heart Rate and Blood Pressure Time Series." Paper presented at 2010 Computing in Cardiology Conference, IEEE, September, pp. 81–84.
32. Sun, J., Sow, D., Hu, J., and Ebadollahi, S. (2010) "A System for Mining Temporal Physiological Data Streams for Advanced Prognostic Decision Support." Paper presented at 2010 IEEE International Conference on Data Mining, IEEE, December, pp. 1061–1066.
33. Cao, H., Eshelman, L., Chbat, N., Nielsen, L., Gross, B., and Saeed, M. (2008) "Predicting ICU Hemodynamic Instability Using Continuous Multiparameter Trends." Paper presented at 2008 30th Annual International Conference of the IEEE Engineering in Medicine and Biology Society, IEEE, August, pp. 3803–3806.
34. Reich, D. L. (2011) *Monitoring in Anesthesia and Perioperative Care*. Cambridge: Cambridge University Press.

35. Smolinska, A., Hauschild, A. C., Fijten, R. R. R., Dallinga, J. W., Baumbach, J., and Van Schooten, F. J. (2014) "Current Breathomics—A Review on Data Pre-Processing Techniques and Machine Learning in Metabolomics Breath Analysis." *Journal of Breath Research*, 8(2): 027105.

36. Le Roux, P., Menon, D. K., Citerio, G., Vespa, P., Bader, M. K., Brophy, G. M., ... and Taccone, F. (2014) "Consensus Summary Statement of the International Multidisciplinary Consensus Conference on Multimodality Monitoring in Neurocritical Care." *Neurocritical Care*, 21(2): 1–26.

37. Tisdall, M. M., and Smith, M. (2007) "Multimodal Monitoring in Traumatic Brain Injury: Current Status and Future Directions." *British Journal of Anaesthesia*, 99(1): 61–67.

38. Hemphill, J. C., Andrews, P., and De Georgia, M. (2011) "Multimodal Monitoring and Neurocritical Care Bioinformatics." *Nature Reviews Neurology*, 7(8): 451–460.

39. Pantelopoulos, A., and Bourbakis, N. G. (2009) "A Survey on Wearable Sensor-Based Systems for Health Monitoring and Prognosis." *IEEE Transactions on Systems, Man, and Cybernetics, Part C (Applications and Reviews)*, 40(1): 1–12.

40. Winkler, S., Schieber, M., Lücke, S., Heinze, P., Schweizer, T., Wegertseder, D., ... and Koehler, F. (2011) "A New Telemonitoring System Intended for Chronic Heart Failure Patients Using Mobile Telephone Technology—Feasibility Study." *International Journal of Cardiology*, 153(1): 55–58.

41. Sow, D., Turaga, D. S., and Schmidt, M. (2013) "Mining of Sensor Data in Healthcare: A Survey." In C. Aggarwal (Ed.), *Managing and Mining Sensor Data*. Cham: Springer, pp. 459–504.

42. Prasad, S., and Sha, M. N. (2013) "NextGen Data Persistence Pattern in Healthcare: Polyglot Persistence." In 2013 Fourth International Conference on Computing, Communications and Networking Technologies (ICCCNT), IEEE, July, pp. 1–8.

43. Brophy, J. T. (1992) "WEDI (Workgroup for Electronic Data Interchange) co-chair predicts big savings from EDI." *Computers in Healthcare* 13(10): 54–57. PMID: 10122897.

Index

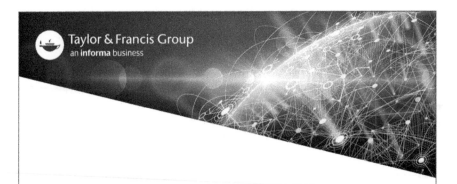

Milton Keynes UK
Ingram Content Group UK Ltd.
UKHW050305161024
449569UK00033B/375